HYDROGEN: ITS TECHNOLOGY AND IMPLICATIONS

Editors

Kenneth E. Cox
Project Manager
Thermochemical Hydrogen
Los Alamos Scientific Laboratory
University of California
Los Alamos, New Mexico

K. D. Williamson, Jr.
Assistant Division Leader
Los Alamos Scientific Laboratory
University of California
Los Alamos, New Mexico

SERIES OUTLINE

Volume I
Hydrogen Production Technology

Volume II
Transmission and Storage of Hydrogen

Volume III
Hydrogen Properties

Volume IV
Utilization of Hydrogen

Volume V
Implications of Hydrogen Energy

Hydrogen: Its Technology and Implications

Volume V
Implications of Hydrogen Energy

Editors:

Kenneth E. Cox

Project Manager
Thermochemical Hydrogen Project
Los Alamos Scientific Laboratory
University of California
Los Alamos, New Mexico

K. D. Williamson, Jr.

Assistant Division Leader
System Analysis and Assessment Division
Los Alamos Scientific Laboratory
University of California
Los Alamos, New Mexico

CRC PRESS, Inc.
Boca Raton, Florida 33431

Library of Congress Cataloging in Publication Data

Main entry under title:

Implications of hydrogen energy.

(Hydrogen, its technology and implications ; v. 5)
Bibliography: p.
Includes index.
1. Hydrogen as fuel. I. Cox, Kenneth E.
II. Williamson, Kenneth D. III. Series.
TP359.H8H9 Vol. 5 665'.81'08s [665'.81] 78-21140
ISBN 0-8493-5125-1

International Standard Book Number 0-8493-5120-0 (Complete Set)
International Standard Book Number 0-8493-5125-1 (Volume V)

Library of Congress Card Number 78-21140
Printed in the United States

PREFACE TO HYDROGEN: ITS TECHNOLOGY AND IMPLICATIONS

The United States, Western Europe, Japan, and several other countries are presently faced with an energy shortage due largely to an imbalance of energy consumption over fossil energy production. This problem was dramatized in October 1973 during the Arab embargo on the shipment of oil to the United States and the resultant large increases in the price of crude oil. This shortage in energy supply was then termed the "energy crisis." It was a clear demonstration of the nation's dependence on imported petroleum and its vulnerability on both political and economic grounds. It is clear that the above problems would worsen in the future unless more attention and effort are directed toward increasing domestic energy production from both depletable and nondepletable sources and reducing energy consumption.

In the short-term, until the year 2000, coal and nuclear energy are expected to play dominant roles in meeting the energy shortage despite the environmental restrictions that hamper the production and consumption of high-sulfur coal and similar difficulties (siting and radioactive waste disposal) that have slowed the development of nuclear energy. In the long-term, beyond the year 2000, it is imperative that all forms of renewable energy be developed. These include solar energy, in such forms as wind, ocean thermal gradients, and biomass; geothermal energy; and fusion.

A major problem with several of the renewable energy sources is that they are intermittent and their energy density is low; thus, there is a need for an energy carrier that can act as both a storage and transportation medium to connect the energy source to the energy consumer. Many of the renewable energy forms, together with coal and fission exhibit their energy in the form of heat release. It is necessary to develop an energy carrier, other than electricity, to supply the transportation sector as well as overcome the problems of electrical storage.

Hydrogen, the lightest element, has been suggested as the energy carrier of the future. In itself, it is not a primary energy source but rather serves as a medium through which a primary energy source (such as nuclear or solar energy) can be stored, transmitted, and utilized to fulfill our energy needs. There are several distinct advantages to the use of hydrogen as an energy medium. It can be made from water, an inexhaustible resource. On combustion, water is the main product; thus, hydrogen can be regarded as a clean, nonpolluting fuel. Indications from current research efforts suggest that hydrogen may be produced from high-temperature heat sources at an efficiency greater than that of electrical generation, thereby making hydrogen a more economical energy source than electricity. Technology has already been developed for storing hydrogen as a pressurized gas, a cryogenic liquid, or in the form of a metal hydride. Systems for transporting hydrogen as a gas or a liquid have been developed with liquid hydrogen playing a major part in NASA's putting a man on the moon. Finally, hydrogen is of value as a chemical intermediate, being used in fertilizer manufacture, methanol synthesis, and petroleum treatment. This area of hydrogen utilization represents 3% of today's energy consumption and is expected to grow by a factor of five by the year 2000. The above concept of using hydrogen is termed the "hydrogen energy economy" and has been receiving an increasing amount of attention from energy scientists and engineers in the United States and abroad.

This series in five volumes represents a serious attempt at providing information on all aspects of hydrogen at the postgraduate and professional level. It discusses recent developments in the science and technology of hydrogen production; hydrogen transmission and storage; hydrogen utilization; and the social, legal, political, environmental, and economic implications of hydrogen's adoption as an energy medium. Although there are several reports of selected studies on hydrogen as a fuel, this is the first comprehensive reference book that covers a wide range of topics of notable interest and timely importance.

Volume I of the series discusses such topics as hydrogen production from fossil fuels, nuclear energy, and solar energy. Hydrogen production technology from water by traditional methods such as water electrolysis and newer attempts to split water thermochemically are included with details of current research efforts and future directions.

Volume II provides detailed design information on systems necessary for the storage, transfer, and transmission of gaseous and liquid hydrogen. Cost factors, technical aspects, and models of hydrogen pipeline systems are included together with a discussion of materials for hydrogen service. Metallic hydride gaseous storage systems for the utility and transportation industry are covered in detail, and the design Dewars and liquid hydrogen transfer systems are examined.

Volume III focuses on hydrogen's properties and provides in one location all of the hydrogen data measured and compiled by the National Bureau of Standards, Cryogenic Division. The properties are individually discussed, and tables of data are provided. The properties of slush hydrogen are also included.

Volume IV covers the present and future uses of hydrogen. Hydrogen has been suggested as a prime candidate for both air and surface transportation. In the utility industry, hydrogen systems for peak shaving promise to play an important future role. Both present and future domestic and industrial applications of hydrogen are surveyed. These include present uses in ammonia and methanol synthesis and future uses in the direct hydrogasification of coal to synthetic natural gas. Important to all of these applications are the safety considerations in the use of hydrogen to allow for public acceptance of hydrogen's role as an energy medium.

Volume V is primarily concerned with the nontechnical aspects of hydrogen. Economics of hydrogen energy systems will play a major part in determining the time frame for hydrogen's adoption. Cost analyses of such systems with return on investment considerations are surveyed from the point of view of production, transmission, and storage of hydrogen. The environmental, political, social, and legal implications of new secondary energy forms such as hydrogen are discussed with reference to governmental energy policy, the social costs of energy production and use, and the public's acceptance of a hydrogen energy medium.

The unusually broad nature of hydrogen demands the expertise that could only be provided by a wide authorship; thus, some of the authors are the original authorities in their respective fields. Although the subject matter treated in each chapter is, in general, the author's research work and his critical review of the state-of-the-art, the authors have had complete freedom in choosing the particular important areas to be emphasized. As a result, some chapters treat the subject matter in more detail than others with a greater emphasis on the engineering or design aspects of a particular system. Therefore, each chapter possesses its own special feature and appealing points. Due to the limited space in the series, the editors have encouraged each author to supply an extensive list of references at the end of his chapter for the benefit of interested readers. Detailed author and subject indexes have been provided at the end of each volume.

The editors, while striving to avoid duplication, have allowed some degree of overlap in certain of the chapters for the sake of continuity and allowing the reader to view a particular topic from two or more points of view. Further volumes on the topic of hydrogen are planned, and we wish to hear from our readers as to areas that might have been neglected or deserve a special chapter on their own.

We would like to express our sincere thanks to these authors and the staff of CRC Press, Inc. in particular Mrs. Gayle Tavens and Miss Sandy Pearlman, for their efforts in making these volumes possible. Lastly, we would like to thank our wives, Patricia R. Cox and Ruth S. Williamson, for their encouragement and help during the time it took to edit these five volumes.

K. E. Cox
K. D. Williamson, Jr.
Los Alamos, New Mexico
March 1975

PREFACE TO VOLUME V: IMPLICATIONS OF HYDROGEN ENERGY

The technology of producing hydrogen has progressed considerably since the discovery of the decomposition of water by an electric current by Nicolson and Carlisle in 1800 followed by Faraday's discovery of the laws governing electrolysis in 1833. Most of today's hydrogen is produced by the steam reforming of methane and the partial oxidation of petroleum. With diminishing fossil fuels, reappraisal of older technologies for producing hydrogen, such as electrolysis, and development of newer methods to produce hydrogen directly from renewable energy sources, such as solar energy and fusion, has begun.

This volume is the result of certain initial evaluations of the legal, and socio-economic factors that enter the energy arena as a transition from a "capital-based" fossil fuel economy to an "income-based" renewable energy economy that uses hydrogen as an energy "medium" takes place.

Chosen for fuller coverage are the economics of hydrogen production, transmission, distribution, and end-use; the effect of a "hydrogen economy" on the environment; a socio-political view of energy with emphasis on hydrogen as the major fuel; and the legal aspects of adopting a new hydrogen energy system. In part, the treatment has been qualitative rather than quantitative as befits the subject material. There is also an element of looking towards the future in terms of "scenarios" as the present energy situation deteriorates as regards fossil fuel reserves. With the rapidity of events occurring in the field of energy at the present, the authors are commended as to the foresight and intuition shown in their respective chapters.

A word of thanks is due all the participants in this volume, the authors who devoted their expertise and time as well as the editorial staff of CRC Press, Ms. Sandy Pearlman and Ms. Barbara Perris, who aided in the preparation of the volume. It is our hope that this volume will prove of use and generate further interest in the fascinating field of hydrogen energy.

Kenneth E. Cox
Los Alamos, New Mexico
December 1978

THE EDITORS

Kenneth E. Cox, Ph.D., is a Staff Member in the High-Temperature Chemistry Group at the Los Alamos Scientific Laboratory, Los Alamos, New Mexico. He is currently performing research in developing practical thermochemical cycles that produce hydrogen by water-splitting. In his previous position as Professor of Chemical Engineering at the University of New Mexico, Albuquerque, he pioneered the development of photovoltaic-electrolytic methods to produce hydrogen from solar energy and water.

Dr. Cox graduated in 1956 from the Imperial College of Science and Technology of the University of London with B.Sc. and A.C.G.I. degrees in chemical engineering. He received his M.A.Sc. from the University of British Columbia in 1959 and his Ph.D. from Montana State University in 1962, also in chemical engineering.

Dr. Cox is a member of the American Institute of Chemical Engineers, the American Chemical Society, and Sigma Xi, the Scientific Research Society. He is also a member of the International Association for Hydrogen Energy and has contributed both as an author and reviewer to the *International Journal of Hydrogen Energy.*

Dr. Cox has published numerous research papers and has given over 100 presentations. His current research interests include: thermochemical generation of hydrogen from water, use of solar energy for hydrogen generation, techno-economic energy evaluations as well as traditional chemical engineering, e.g., thermodynamics of separation processes, process design, and dispersion phenomena in fluids.

K. D. Williamson, Jr., Ph.D., is Assistant Division Leader of the Systems Analysis and Assessment Division, Los Alamos Scientific Laboratory, Los Alamos, New Mexico.

Dr. Williamson received his B.S., M.S., and Ph.D. degrees in chemical engineering from Pennsylvania State University in 1957, 1959, and 1961, respectively.

Dr. Williamson is a member and lecturer for the New Mexico Academy of Sciences and has served as an invited lecturer at the University of Tennessee and New Mexico Highlands University.

Dr. Williamson has published more than 35 research papers. His current research interests include technology assessments and systems analyses of energy-related topics in support of research and development decision makers at Los Alamos and the State and National government levels.

CONTRIBUTORS

John R. Bartlit, Ph.D.
Staff Member
Los Alamos Scientific Laboratory
University of California
Los Alamos, New Mexico

Thomas C. Cady
Professor
College of Law
West Virginia University
Morgantown, West Virginia

Jesse Hord, Ph.D.
Cryogenics Division
Institute for Basic Standards
U.S. Department of Commerce
National Bureau of Standards
Boulder, Colorado

Jack D. Salmon, Ph.D.
Department of Political Science
Omega College
University of Western Florida
Pensacola, Florida

TABLE OF CONTENTS

Chapter 1

Economics of Hydrogen

Chapter 1
ECONOMICS OF HYDROGEN

J. Hord and W. R. Parrish

TABLE OF CONTENTS

1.1. INTRODUCTION

Although some familiarity with economic principles is assumed, this chapter is intended to be useful to the reader who has no special training in economics. It is not an exercise in economic analysis; rather, we endeavor to provide the cost data necessary for such analyses and to make possible economic comparisons with competing fuels or feedstocks. Examples of simple economic analyses are given. The major thrust of this chapter will be directed toward determining the cost parameters needed to assess the cost of producing, transmitting, storing, and using hydrogen in hydrogen energy systems. Technical details concerning production, transmission, storage, etc. are treated elsewhere in this series and are only mentioned in this chapter to the extent necessary to specify relevant cost parameters. Where possible (and practical), cost parameters are presented in a manner that permits updating to account for increased costs of energy, materials, labor, etc. In the data presented herein, the heat of combustion of hydrogen is always evaluated at the higher heating value (HHV).

Cost analysis is intimately and inextricably interwoven with energy system design. Using a specific costing procedure, we can design a system to provide minimum capital investment, minimum annual operating charges, or minimum capital investment plus annualized operating charges. Each of these costing options can result in different system designs, as can variations in the costing procedure. Conversely, system characteristics strongly influence the accuracy and validity of the cost analysis — no cost analysis can be assigned credibility unless due consideration has been given to the design of the energy system. Thus, cost analysis and total system design go hand in hand and must be handled simultaneously and with equal skill. The strong dependence of hydrogen product cost on total system analysis will be apparent throughout various sections of this chapter.

1.2. FUNDAMENTALS AND TERMINOLOGY OF ECONOMICS

The economic terms and basic economic principles used in this chapter are described below. It is emphasized that all costs presented herein, unless specifically excepted and identified, are adjusted to September 1975 in accordance with the inflation index curve for chemical process plants.[1] This cost inflation index was judged most appropriate for the hydrogen systems under consideration and is shown in Figure 1.

We consider three of the most common economic analyses: rate of return on investment (RRI), discounted cash flow rate of return (DCF), and utility financing method (UFM). For details of the first two methods, along with other analyses, see any process economics text, e.g., Stermole[2] and Happel.[3] The UFM is described by Siegel et al.[4] The RRI method is sometimes called the ROI or ROR method, and the DCF method is sometimes referred to as the DCFROR method of economic analysis.

The simplest method of economic analysis is the annual rate of return method given by

$$RRI = p/(I + I_w) \tag{1}$$

In Equation 1, I is the total plant investment including interest paid during construction, and I_w is the working capital, i.e., the money tied up in raw material, product inventories, and accounts receivable, and other cash required to operate the facility; p represents the annual net profit after taxes and depreciation. We may also write

$$p = R - e \cdot I - t\,(R - d \cdot I) \tag{2}$$

where e and d are the annual depreciation rates for accounting and tax purposes, respectively, t is the combined federal and state income tax rate, and R is the gross annual profit. The latter is the gross return minus all operating costs; it excludes depreciation and state and federal income taxes, but includes feedstocks, utilities, operating and maintenance (O & M) costs, property taxes and insurance (T & I) costs.

The depreciation terms e and d can be computed in several ways and are frequently used interchangeably. One of the most common methods used to depreciate the investment for both accounting and tax purposes is straight line depreciation. In this case, e is the reciprocal of n, the life of the project in years. To obtain tax advantages, the plant can be depreciated more rapidly by using an accelerated depreciation technique. The most commonly used method of accelerated depreciation is the double declining balance which requires that

$$d_j = \left(1 - \frac{2}{n}\right)^{j-1} \left(\frac{2}{n}\right) \tag{3}$$

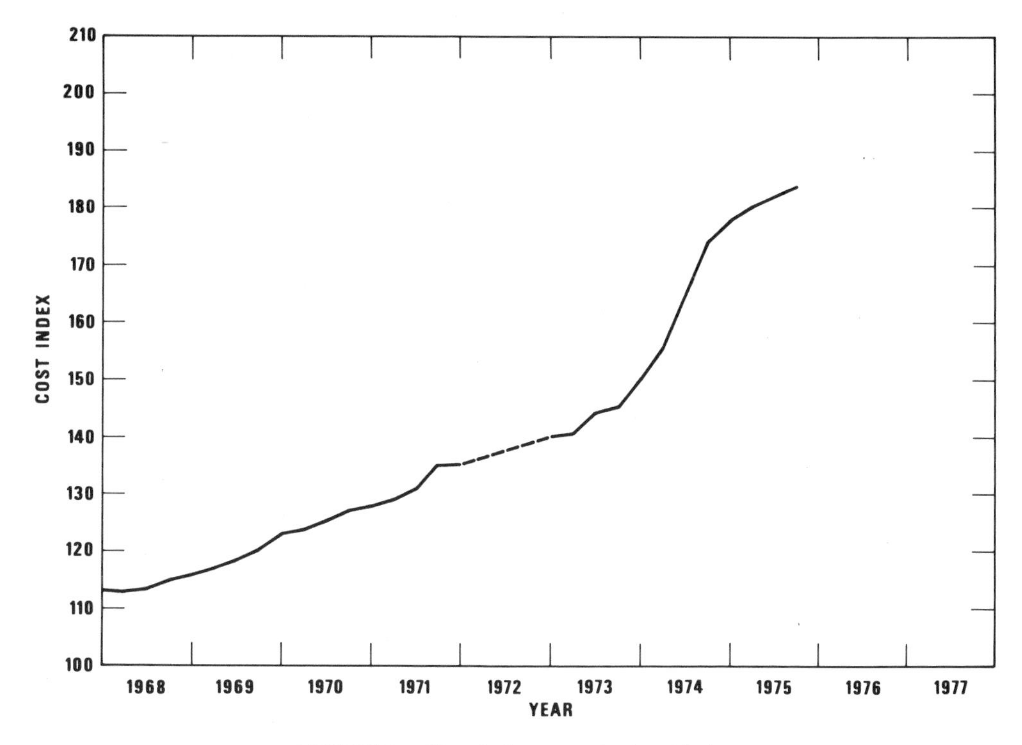

FIGURE 1. Chemical process plant cost index.

where d_j is the depreciation in the jth year of operation. This method is sometimes used in the early years of plant operation; straight line depreciation is used in later years to obtain maximum tax deductions. The sum-of-the-digits method requires that

$$d_j = \frac{2(n - j + 1)}{n(n + 1)} \tag{4}$$

Another way to calculate e is to require that a certain fraction of the initial investment be invested each year of the project life into an account yielding an annual interest rate, i. Then

$$e = \frac{i}{(1 + i)^n - 1} \tag{5}$$

where n is the life of the project in years. After putting in n equal "sinking fund" payments of e·I, the total initial investment will be recovered. This accounting procedure is nearly obsolete in modern industry; however, the sinking fund concept is still used to account for the present worth of a capital investment (see Equation 9).

The RRI method is simply a means of specifying that the net profit be some fraction of the total investment. However, it does not consider the present worth of money, i.e., it does not discount future returns to the present. To account for this shortcoming, the venture worth concept is used. This method requires that both the minimum acceptable rate of return, due to the risk of the project, and the company's normal rate of return on its invested capital must be specified. The discounted cash flow (DCF) method sets these two rates of return equal to a common return, i, which will make the venture worth = 0.

The DCF return is obtained from the following equation:

$$0 = \sum_{j=1}^{n} \frac{(1 - t)}{(1 + i)^j} R_j + \sum_{j=1}^{n} \frac{d_j(t)I}{(1 + i)^j} - I - \left[\frac{(1 + i)^n - 1}{(1 + i)^n}\right] I_w + \frac{(1 - t)}{(1 + i)^n} S \tag{6}$$

In Equation 6, the first term is the sum of the discounted gross annual returns before income taxes; the second term represents total discounted tax credit due to depreciation; the third term is the initial investment; the fourth term is the present worth of the working capital; and the last term represents the present worth of salvage. S is the salvage value and is usually taken as zero at the end of the project life. If the annual gross return is constant over n years, then

$$(1 - t)\, R \sum_{j=1}^{n} \frac{1}{(1+i)^j} = (1 - t)\, R \left[\frac{(1+i)^n - 1}{i\,(1+i)^n}\right] \quad (7)$$

The depreciation term becomes

$$\sum_{j=1}^{n} \frac{d_j(t)I}{(1+i)^j} = \frac{t \cdot I}{n} \left[\frac{(1+i)^n - 1}{i\,(1+i)^n}\right] \quad (8a)$$

or

$$t \cdot I \sum_{j=1}^{n} \frac{d_j}{(1+i)^j} = \frac{2t \cdot I}{n(n+1)i} \left[n - \frac{(1+i)^n - 1}{i\,(1+i)^n}\right] \quad (8b)$$

or

$$t \cdot I \sum_{j=1}^{n} \frac{d_j}{(1+i)^j} = \frac{2t \cdot I}{n} \left\{\frac{1 - [(1 - 2/n)/(1+i)]^n}{i + 2/n}\right\} \quad (8c)$$

for straight line, sum-of-the-digits, and double declining balance methods, respectively. If the gross return (R) can be estimated, Equations 6, 7, and 8 can then be solved for i, the effective interest rate of return (DCF rate of return). If R_j is not constant but is known, i must be determined by iteration using Equation 6; however, the more common method is to specify i and solve for R, assuming R is constant over n years. This method can be used to establish a product selling price because R is specified.

Utility financing method[4] (UFM) is a more recent procedure used to estimate the selling price of a synthetic fuel. Tables 1 and 2 give the necessary equations and estimating procedures for determining total capital investment. Although this analysis is not used in this chapter, it is frequently used in the literature and is referred to as "utility financing." "Industrial financing" normally refers to the DCF method with a specified rate of return, since most large businesses

TABLE 1

Average Gas Cost Equation (UFM)

Basis:

20-year project life
5%/year straight-line depreciation on total capital investment, excluding working capital
48% federal income tax rate

Definition of terms:

^aI	=	total capital investment, 10^6 \$
I_w	=	working capital, 10^6 \$
N	=	total net operating cost in first year, 10^6 \$/year
G	=	annual gas production, 10^{12} Btu/year
f	=	fraction debt
i	=	interest on debt, %/year
r	=	return on equity, %/year
p	=	return on rate base, %/year

Equation for return on rate base:

$$p = (f)i + (1 - f)\, r$$

General gas cost equation:

$$\text{Average Gas Cost, \$/}10^6\text{ Btu} = \frac{aN + 0.05\,(I - I_w) + 0.005\,[p + \frac{48}{52}(1 - f)r]\,(I + I_w)}{G}$$

Values of the parameter a:

Operating cost approach	Plant startup date	Value of a
Without inflation during project life	All years	1.0000
With inflation during project life	1975	1.3726
	1980 and beyond	1.3435

[a]See Table 2.

From Siegel, H. M., Kalina, T., and Marshall, H. A., paper presented to the Federal Power Commission, Washington, D.C., June 12, 1972.

TABLE 2

Basis for Calculating Total Capital Investment[a] (UFM)

Total plant investment	
All onsites plant sections	XXX
All utilities and offsites (Including fresh-water treating, cooling towers, power generation and distribution, steam generation, pollution control facilities, site preparation, offices, shops, control houses, etc.)	XXX
Contractor's overhead and profit	XXX
Engineering and design costs	XXX
Subtotal plant investment	XXX
Project contingency (15% of subtotal plant investment)	XXX
Development contingency[b] (7% of subtotal plant investment)	XXX
Total plant investment	XXX
Total capital requirement	
Interest during construction (interest rate × total plant investment × 1.875 years average period)	XXX
Startup costs (20% of total gross operating cost)[c]	XXX
Working capital (sum of: raw materials inventory of 60 days at full rate, materials and supplies at 0.9% of total plant investment, and net receivables at 1/24 of annual gas revenue @ $\$1.00/10^6$ Btu)	XXX
Total capital requirement	XXX

[a]All items in parentheses refer to particular bases used by the Synthetic Gas-Coal Task Force.
[b]Not required for processes already developed.
[c]Based on capitalization of 40% of the full-rate gross operating costs during a 6-month startup period. (Assumes that 60% of the costs during the startup period are covered by revenue from gas deliveries.)

use this method. Small and medium-sized businesses may use the RRI method or a "payout time" method of economic analysis. The latter is a simple modification of the RRI method, and provides an estimate of the time required to recover the capital investment.

Using the RRI method, we can develop a simplified procedure for performing preliminary economic analyses. The sinking fund concept as expressed by Equation 5 does not consider the time-appreciated value of the initial investment; i.e., if I dollars are placed at an interest rate (i), the value of the account at the end of n years will be $I\ (1 + i)^n$ dollars if the interest is compounded annually. If the sinking fund deposit factor, Equation 5, is modified to account for the present worth of a capital investment, the capital recovery factor (CRF) is obtained,

$$CRF = \frac{i\ (1 + i)^n}{(1 + i)^n - 1} \qquad (9)$$

CRF·I is the annual deposit required to provide a sum of money = $I\ (1 + i)^n$ at the end of n years, where the interest at rate i is compounded annually and paid on the accumulated deposit at the end of each year. CRF·I may also be used to designate the annuity required to meet annual mortgage payments on borrowed capital.

Frequently the CRF is combined with property taxes and insurance (T & I) to form the fixed charge rate (FCR). The FCR is an annual charge rate that is independent of the plant factor – the fraction of time, on an annual basis, that the plant is operating at design throughput. Strictly speaking, O & M costs are not fixed charges. However, for many preliminary economic analyses they can be considered as a constant fraction of the initial capital investment. Combining FCR and O & M with a specified incremental rate of return on the investment (IRRI) before income taxes gives the annual charge rate (ACR). Then, ACR·I is the annual cost of owning and operating the plant, excluding feedstock and utility costs. Adding ACR·I to utility and feedstock costs gives the total annual plant charges and also the total annual value of the product; division by the product throughput yields the unit selling price.

A series of equations may be developed to illustrate the foregoing discussion. Combining Equations 1 and 2 and setting e = CRF we obtain an expression for the incremental rate of return on investment,

$$(I + I_W)\ IRRI' = R - CRF \cdot I - t\,(R - d \cdot I) \qquad (10)$$

IRRI′ may be viewed as an additional rate of return above that embodied in the CRF. To simplify matters we let $R = ACR \cdot I - [(T\ \&\ I) + (O\ \&\ M)]\ I$ and note that $(I + I_W) \approx I$. Then Equation 10 may be rewritten as

$$IRRI' = ACR - [(T\&I) + (O\&M)] - CRF - t\,(R - d \cdot I)/I \qquad (11)$$

Rearranging Equation 11, we obtain expressions for the annual charge rate

$$ACR = [CRF + (T\&I)] + (O\&M) + [IRRI' + t\,(R - d \cdot I)/I] \qquad (12a)$$

and

$$ACR = FCR + (O\&M) + IRRI \qquad (12b)$$

where IRRI is the incremental rate of return, before income taxes, on the initial capital investment. In a simplified economic analysis where feedstock and utility costs are also incurred

$$[(ACR \cdot I) + (\text{annual feedstock and utility costs})] \div \text{throughput} = \text{unit cost of product.}$$

The ACR can be constructed in many different ways. It may be presented as the sum of percentages of I that are allocated to CRF, T & I, O & M, and a state and federal income tax annuity (ITA). It can also be derived by summing the percentages of I that are attributable to straight-line depreciation, T & I, O & M, ITA, and a specified RRI. Similarly, the FCR lacks a common definition: it sometimes includes (1) CRF and T & I (as in this chapter); (2) CRF, T & I, and ITA; (3) CRF, T & I, ITA, and O & M; and (4) depreciation, T & I, O & M, and RRI. Thus, it is apparent that the definition of FCR is sometimes equivalent to that of the ACR (as defined herein). Study of Equations 10 through 12 will also show that the IRRI includes an allowance for ITA.

1.3. HYDROGEN PRODUCTION COSTS

This section considers the cost of producing hydrogen from fossil fuels and water; it also includes costs associated with converting hydrogen at 14.7 psia (0.1 MPa) and 80°F (26.7°C) to liquid, slush, solid, or high-pressure gas. A few potential cost credits that could lower the overall cost of hydrogen are also described herein.

1.3.1. Hydrogen Production

There are many processes for producing hydrogen from fossil fuels and/or water. However, we will consider only existing and near-term technology so that reasonable cost estimates can be made.

Two-stage reactions are required to produce hydrogen from fossil fuels. The first stage is the reaction of the fossil fuel with steam and/or oxygen to produce a synthesis gas (syngas) containing hydrogen, carbon monoxide, carbon dioxide, and water; this step requires heat. The next stage shifts carbon monoxide and steam to more hydrogen and carbon dioxide. This shift conversion reaction is exothermic, and the recovered heat is used in the first stage. Table 3 gives estimated capital costs[5] for producing 2440 $\times 10^6$ Btu/h (18,135 kg/h) of hydrogen from fuel oil and coal and by steam reforming and partial oxidation of methane. Table 4 gives estimated annual operating costs and working capital for each process. These cost data are provided by the same economic evaluations group;[5] therefore, an internal consistency within the cost estimates for each process is assured. All costs were inflated to September 1975 dollars. It is estimated that capital costs vary according to the 0.8 power of the plant capacity.

Figures 2 and 3 show the estimated production cost and selling price of hydrogen for the various processes as a function of the cost of the primary fossil fuel feedstock. The solid portion of each line denotes the present to near-term range of fossil fuel costs. The production costs are based on an annual fixed charge rate of 13%; this includes 2% for local taxes and insurance and 11% for retirement of debt (assuming a 9% loan for 20 years). Also, an annual plant factor of 90% is assumed.

The selling prices are based on the following assumptions: a) a discounted cash flow rate, before taxes, of 20%; b) a federal income tax rate

TABLE 3

Capital Costs for Producing 2440 × 10⁶ Btu/h (18,135 kg/h) of Hydrogen from Fossil Fuels

Process	Steam reforming	Partial oxidation		Coal gasification
Feedstock	Methane	Methane	Fuel oil	Coal
Product purity, mol %	96.9	96.6	97.1	97.3
Product pressure, psia (MPa)	465 (3.2)	465 (3.2)	465 (3.2)	465 (3.2)
Thermal efficiency	72.6	66.9	58.5	51.5
Total plant investment (TPI) 10^6 $	25.3	46.4	67.7	63.6
Unit costs (% of TPI)				
Coal prep	–	–	–	3.4
Reactor	56.3	2.6	5.4[a]	2.9[b]
Heat recovery (stages)	7.6 (3)	2.6 (3)	8.2 (2)	5.3 (2)
Shift converters (stages)	2.2 (1)	2.2 (2)	1.9 (2)	1.8 (1)
Purification (stages)	14.2 (1)	12.4 (2)	17.5 (2)	24.4 (2)
Product compression	2.4	2.1	1.7	1.4
Oxygen plant	–	40.3	28.6	31.5
Sulfur recovery	–	–	–	1.4
Steam plant	–	21.1	20.2	11.2
Plant facilities	6.2	6.2	6.3	6.3
Plant utilities	8.9	9.0	9.0	9.0
Catalysts	2.2	1.5	1.2	1.4
	100.0	100.0	100.0	100.0
9% interest during construction, % of TPI	2.5	2.5	2.5	2.5

[a]Cost includes a waste heat recovery unit.
[b]Cost includes dust removal.

Adapted from Katell, S., Bureau of Mines, U.S. Dept. of the Interior, Morgantown, W. Va., private communication, 1973.

of 48%; c) a plant factor of 90%; d) a depreciation rate based on the sum-of-the-digits for 16 years; and e) a plant life of 20 years and no salvage value at the end of 20 years. Land costs and state income taxes have been disregarded in both figures.

Curve 4 on Figures 2 and 3 shows that methane is the most economical source of hydrogen as long as methane is available at less than $1.00 per 10^6 Btu. If the cost of coal does not escalate too rapidly, coal gasification may be a more economical source of hydrogen as methane costs increase – compare Curves 1 and 4 on Figures 2 and 3. It appears that fuel oil is much too expensive to be used as a source of hydrogen.

If there is a need for low-Btu gas, e.g., in power generation as well as hydrogen, the steam-metal oxide process becomes attractive on an economic and efficiency basis.[6-8] The first step of this process produces low-Btu syngas via coal gasification. Instead of shifting the syngas, it is used to reduce oxides of iron or tin. The metal is then reacted with steam to form hydrogen and the metal oxide. Even after the reduction step, the syngas still contains enough hydrogen and carbon monoxide to be a useful low-Btu fuel. This process eliminates the need for oxygen plants, shift converters, and purification units. This process is not attractive unless the low-Btu syngas can be used.

The most abundant source of hydrogen is water. Currently, electrolysis is the only commercial process for producing hydrogen directly from water. Table 5 gives estimated cost data for a large, commercial electrolyzer; unit capital costs for small electrolyzers (0.06 × 10^6 Btu/h or 0.45 kg/h and smaller) are six to ten times greater[10]

TABLE 4

Annual Operating Cost and Working Capital for Producing 2440 × 10⁶ Btu/h (18,135 kg/h) of Hydrogen from Fossil Fuels

Process	Steam reforming	Partial oxidation		Coal gasification
Feedstock	Methane	Methane	Fuel oil	Coal
Estimated annual operating cost,[a] 10^6 $/year				
Direct costs:				
Methane (@ $1.00/$10^6$ Btu)	26.25	22.51	–	–
Fuel Oil (@ $2.00/$10^6$ Btu)	–	–	43.90	–
Coal ($40.00 /ton)	–	9.52	17.44	57.04
Power (@ $0.025/kWh)	1.08	–	–	0.65
Raw water (@ $0.10/$10^3$ gal)	0.17	0.13	0.20	0.24
Chemicals and catalysts	1.62	0.18	0.35	0.34
	29.12	32.34	61.89	58.27
Direct labor[b] (@ $6.00/h)	1.20	2.12	2.12	1.21
Plant maintenance labor[b] (@ $6.00/h)	0.39	1.11	1.63	2.05
Maintenance materials	0.21	0.40	0.61	0.76
Payroll overhead[c]	0.40	0.81	0.94	0.81
Operating supplies[d]	0.12	0.30	0.45	0.56
Total direct costs	31.44	37.08	67.64	63.66
Indirect costs[e]	0.77	1.57	1.92	1.83
Total operating cost excluding fixed costs	32.21	38.65	69.56	65.49
Sulfur credit (@ $18/ton)	–	–	–	0.31
Operating cost after credit (excluding fixed charges)	32.21	38.65	69.56	65.18
Estimated working capital,[f] $$10^6$	1.62	2.48	3.48	2.43

[a]Costs based on 90% stream factor.
[b]Labor costs include supervisor costs.
[c]Payroll overhead estimated at 25% of payroll.
[d]Operating supplies estimated at 20% of total maintenance cost.
[e]Indirect costs include administration and general overhead; taken at 40% of labor, maintenance and supplies.
[f]Working capital estimated to be 25% of annual cost excluding feedstock, depreciation and sulfur credit.

Adapted from Katell, S., Bureau of Mines, U.S. Dept. of the Interior, Morgantown, W.Va., private communication, 1973.

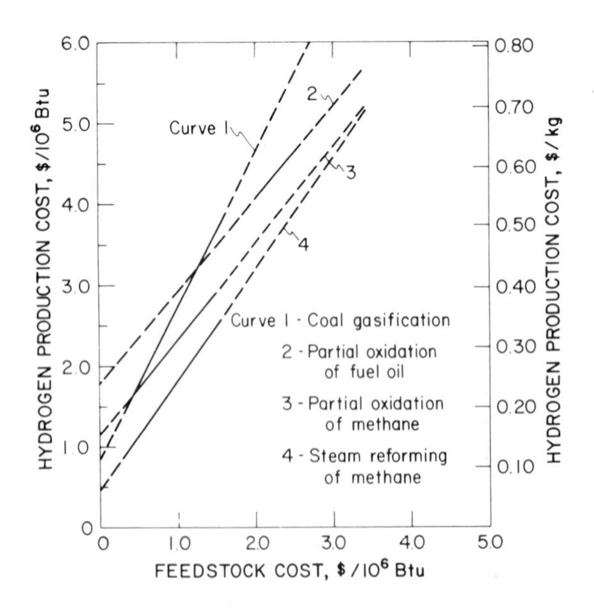

FIGURE 2. Estimated cost of producing hydrogen from fossil fuels (as a function of feedstock costs).

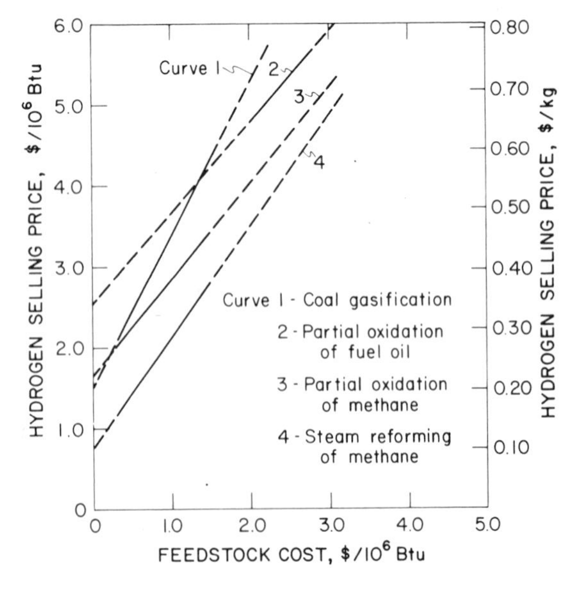

FIGURE 3. Estimated selling price of hydrogen produced from fossil fuels (as a function of feedstock costs).

TABLE 5

Costs of Large, Water Electrolysis Unit[a]

Capital cost of electrolyzer	$\$260/(kW_t)_{H_2}$ (76,147 $\frac{\$\text{-h}}{10^6\ \text{Btu}}$ or 10,245 $\frac{\$\text{-h}}{\text{kg}}$)
a.c./d.c. Converter	$\$45/kW_e$ (17,572[a] $\frac{\$\text{-h}}{10^6\ \text{Btu}}$ or 2,364[a] $\frac{\$\text{-h}}{\text{kg}}$)
Estimated installation cost	50% of capital cost
Estimated operating and maintenance	5% of installed capital cost per year
Power consumption[a]	390 kWh/10^6 Btu (52.5 kWh/kg)

[a]Assuming an overall efficiency of 75% (HHV of H_2 output) ÷ (input electrical power). From Electrical Power Research Institute Report EPRI 320-1, Palo Alto, Cal., August 1975).

than those given in Table 5. Figures 4 and 5 give the production cost and selling price of electrolytic hydrogen as a function of power cost and plant factor; these figures rely on the same assumptions used to derive Figures 2 and 3. The solid portion of the 30% plant-factor line defines the range of current and near-term costs of off-peak electrical power (5 to 15 mills/kWh); the solid portions of the 50 and 90% plant-factor lines define the range of current and near-term costs for baseload, busbar electricity (15 to 25 mills/kWh).

Although electrolysis is an established commercial technology, large commercial units have not been emphasized. Most electrolyzer manufacturers feel that capital costs for large units can be reduced by at least 50% and that overall efficiencies can be increased to 80 or 85% by 1980.[10,12] These improvements would lower the cost of electrolytic hydrogen appreciably.

Many thermochemical processes for producing hydrogen from water are being investigated, but little is known about the economics of such systems. Some preliminary, and perhaps speculative, cost estimates were prepared in a recent study[13] that examined a hybrid thermochemical-electrolysis cycle. The cycle consists of two steps:

$$2H_2O + SO_2 \rightarrow H_2 + H_2SO_4 \quad (13)$$

$$H_2SO_4 \rightarrow H_2O + SO_2 + \frac{1}{2}O_2 \quad (14)$$

The first reaction is performed via electrolysis; however, the theoretical energy requirement for this reaction is only 15% of that required for water electrolysis. The decomposition of sulfuric acid takes place at 1600°F (871°C) and will require the development of a very-high-temperature reactor (VHTR). A detailed analysis[13] of this cycle shows that the overall thermal efficiency is about 47%. This is higher than the most optimistic efficiency estimates for nuclear heat to electricity to electrolysis systems. Table 6 provides cost comparisons for hydrogen produced by dedicated nuclear-electrolysis, coal gasification, and thermochemical processes. The electrolysis and coal gasification costs are based on current technology, and all costs are expressed in 1974 dollars; these production costs are based on utility financing, and no attempt was made to adjust these costs to September 1975, as they are presented for comparative purposes only. For more details, the reader is referred to Farbman.[13]

1.3.2. Costs of Compressing, Liquefying, and Solidifying Hydrogen

The most practical compressors for hydrogen service are reciprocating compressors. Table 7 gives estimated cost data for hydrogen compressors as a function of input power. The input power, based on a single-stage adiabatic compression is

$$P = \frac{\dot{m}RT}{\eta}\left(\frac{\gamma}{\gamma - 1}\right)\left[\left(\frac{P_2}{P_1}\right)^{\frac{\gamma}{\gamma - 1}} - 1\right]$$

where $\dot{m}$ = mass flow rate, R = universal gas constant; T = absolute temperature of inlet gas; P_1, P_2 = inlet and exit pressure, respectively; η = overall adiabatic efficiency; γ = ratio of specific heat capacities, 1.41 for hydrogen; P = power; and any consistent system of units may be used.

The maximum overall adiabatic compressor efficiency is currently 82%.[14] Because the maxi-

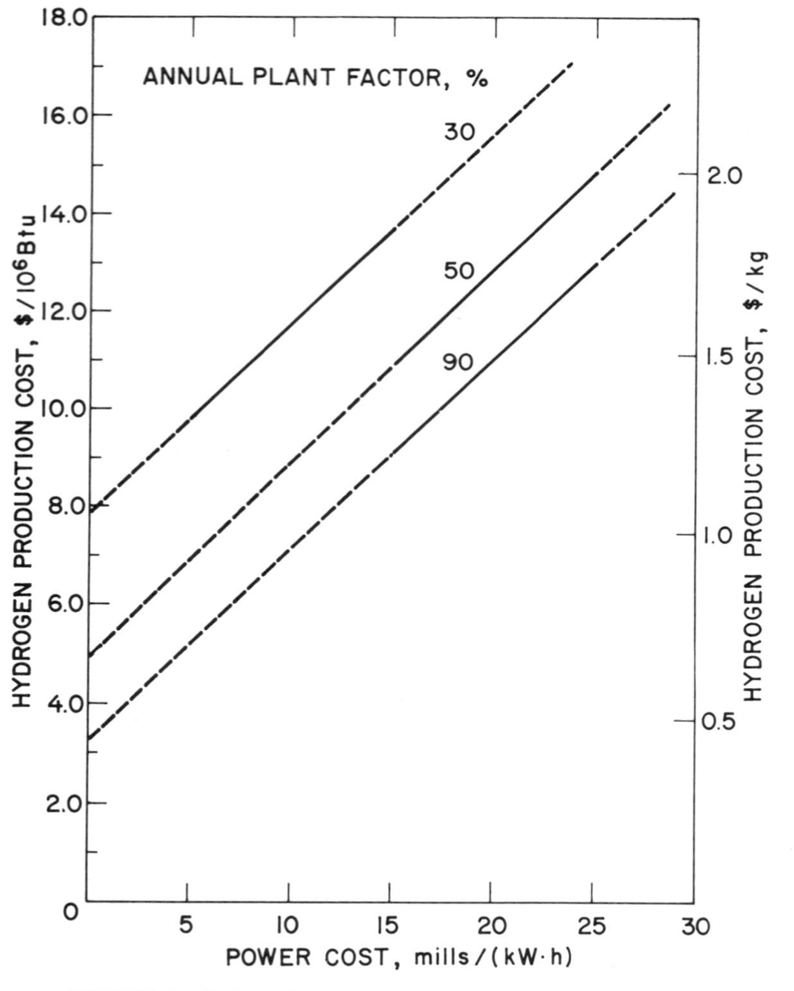

FIGURE 4. Estimated cost of producing electrolytic hydrogen.

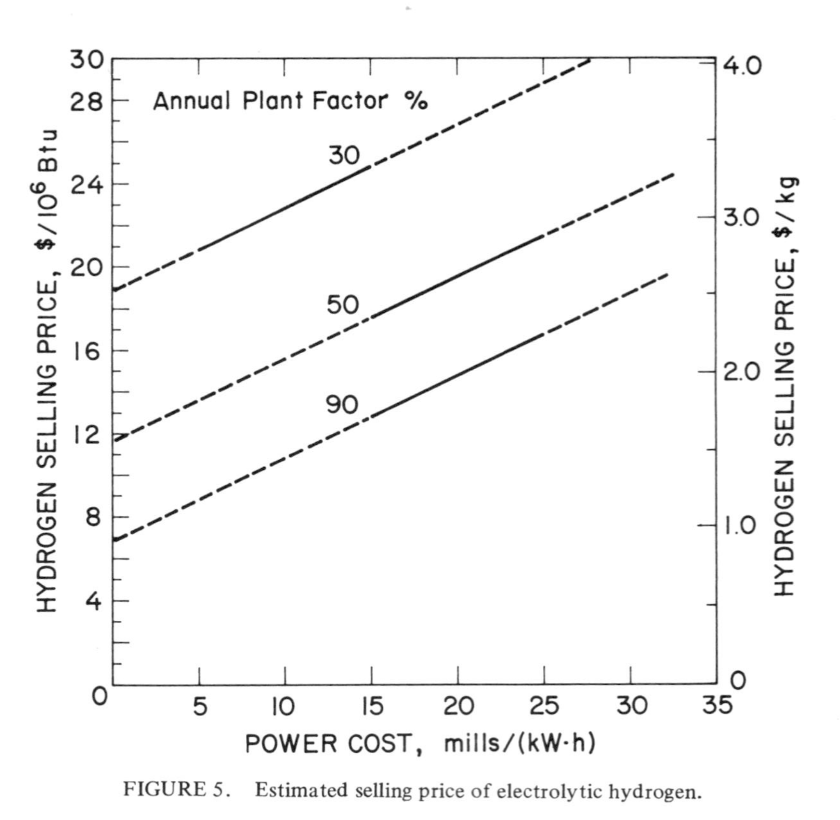

FIGURE 5. Estimated selling price of electrolytic hydrogen.

TABLE 6

Comparison of Hydrogen Production Costs (1974 dollars)

	Electrolysis	Coal gasification	Thermochemical
Thermal efficiency, %	27.6	49.0	46.7
Total plant investment, 10^6 $	505.4	407.3	900.9[a]
Annual production rate,[b] 10^6 Btu (kg)	3.33×10^7 (2.47×10^8)	3.39×10^7 (2.52×10^8)	3.60×10^7 (2.67×10^8)
Estimated annual operating costs, 10^6 $/year			
Fixed charges	75.77	61.04	135.09
O&M	2.25	10.51	4.99
Energy	153.98[c]	67.99[d]	20.07[e]
Total annual costs	232.00	139.54	160.15
Hydrogen production cost, $/$10^6$ Btu ($/kg)	6.98 (0.94)	4.12 (0.55)	4.45 (0.60)

[a]Includes nuclear reactor and thermochemical plant.
[b]Based on an 80% plant factor.
[c]Assumed power cost of 12.8 mills/kWh.
[d]Assumed power cost of 20 mills/kWh, coal cost assumed to be $20/ton.
[e]Assumed nuclear fuel cost of $0.2475/GJ (0.261 $/$10^6$ Btu).

From Farbman, G. H., Report No. NASA-CR-134918, January 1976.

TABLE 7

Cost of Reciprocating Hydrogen Compressors

Size, HP (kW)	200–400 (149–298)	400–700 (298–522)	700–1000 (522–746)	1000–2500 (746–1860)
Cost of compressor[a] (frame, piping, intercoolers and baffles), $/HP[b] ($/kW)	350 (470)	360 (480)	300 (400)	260–270 (350–360)
Cost of electric drive, $/HP ($/kW)	75 (100)	75 (100)	60 (80)	40 (54)

Note: Estimated installation cost is 50% of equipment cost; O&M cost is estimated at 5% of capital cost per year.

[a]Average costs based on actual purchases during 1975; the costs are valid for both single and multiple stages.
[b]Power is based on an overall adiabatic efficiency of 82%.

From Gimbrone, G., Worthington Compressor Company, Buffalo, N.Y., private communication, 1976.

mum temperature of gas leaving a single compression stage is about 300°F (148.9°C), the maximum pressure ratio per stage for hydrogen compression is roughly 3 to 3.2; this pressure ratio depends upon the gas inlet temperature to that stage. Multistage power requirements can be calculated by using the above equation and by assuming the same pressure ratio for each stage.

Cost estimates for producing liquid, slush, or solid hydrogen can be based on power input. Table 8 gives the *ideal* work required to produce liquid, slush, and solid from hydrogen gas at 14.7 psia (0.1 MPa) and 80°F (26.7°C). These calculations are based on the thermodynamic availability change of hydrogen,

$$a_L - a_O = (h_L - h_O) - T_O (s_L - s_O)$$

where a, h, and s are the specific availability, enthalpy, and entropy, respectively. The subscripts O and L denote ambient- and low-temperature

TABLE 8

Ideal Work Required to Produce Liquid, Slush and Solid Hydrogen from Gas at 14.7 psia (0.1 MPa) and 80°F (26.7°C)

	Ideal work required		
Product	kWh/lb	kWh/10^6 Btu	kWh/kg
Saturated liquid @ 14.7 psia (0.1 MPa)	1.801	29.51	3.971
Slush hydrogen: 50-50 mixture of solid and liquid at triple point temperature and pressure	1.985	32.52	4.375
Solid hydrogen at triple point temperature	2.061	33.77	4.543

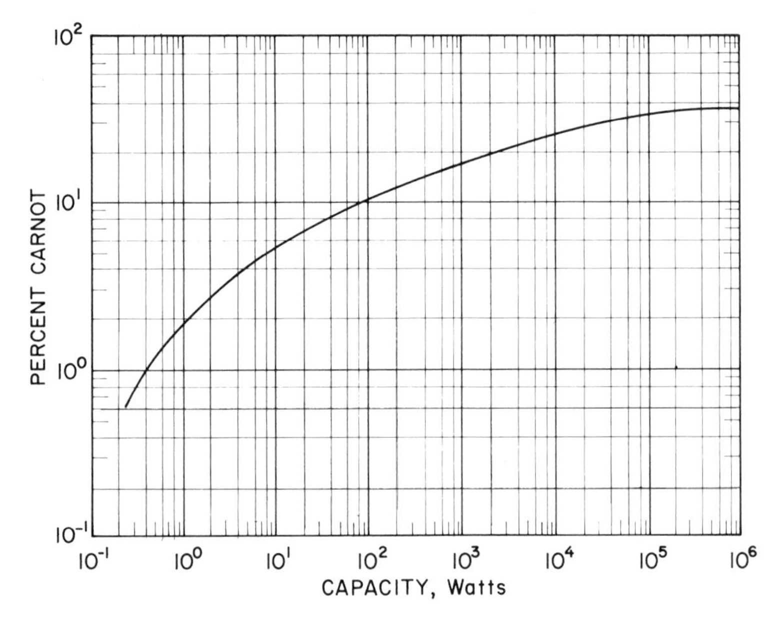

FIGURE 6. Carnot efficiency of hydrogen liquefiers and refrigerators as a function of refrigeration capacity.

conditions, respectively, and T_O is the ambient temperature. Figure 6, which is taken from a survey by Strobridge,[15] gives refrigerator efficiencies as a function of refrigerating capacity. The percent of Carnot efficiency, as defined for refrigerators and liquefiers, is the ideal Carnot work required to remove a specific amount of heat divided by the actual amount of work required to remove the same amount of heat, i.e.,

$$\text{Percent of Carnot Efficiency} = \frac{W_{ideal}}{W_{actual}} = \frac{T_O - T_L}{T_L \cdot W_{actual}}$$

where T is the absolute temperature and the subscripts have the same meaning as before.

The refrigeration capacity is related to the liquefaction rate by

$$\text{Capacity} = \frac{\dot{m}(W_i)T_L}{T_O - T_L}$$

where $\dot{m}$ is the mass flow rate and W_i is the ideal work of liquefaction based on availability calculations. A study by Voth and Daney[16] shows that the maximum efficiency of hydrogen liquefiers is about 36%; this efficiency is limited by the efficiency of the liquefier components (compressors and expansion engines). For liquefiers with capacities of 46.7 × 10^6 Btu/h (347 kg/h) or greater, the liquefier efficiency can be assumed to be 36%.

Figure 7 shows the installed capital cost of hydrogen liquefiers as a function of the actual input power. Operating and maintenance costs are

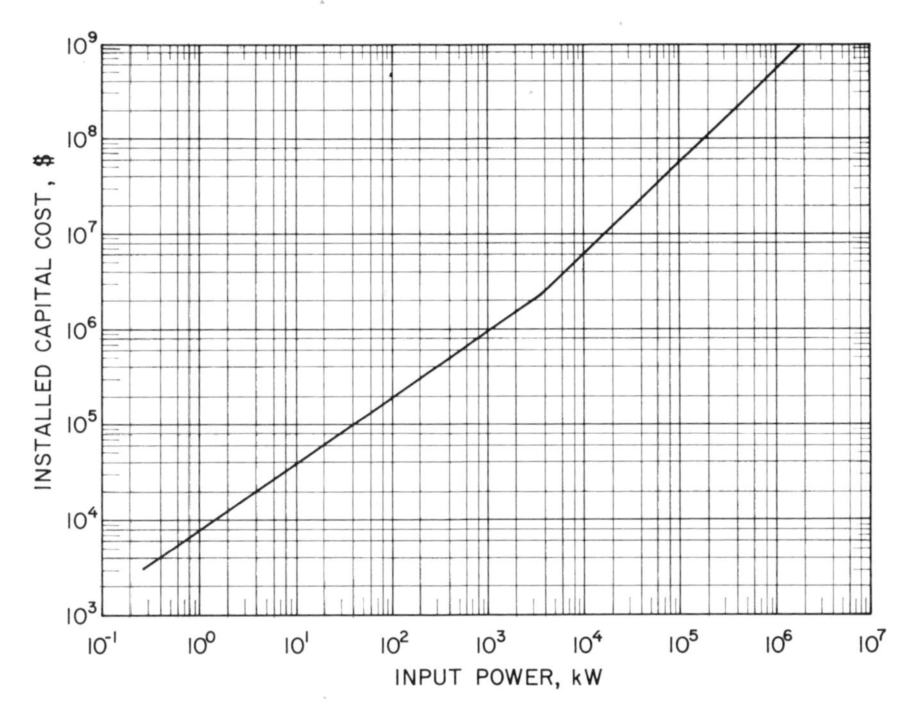

FIGURE 7. Installed capital cost of hydrogen liquefiers and refrigerators.

taken at 5% of the installed capital costs per year.[17] Assuming a Carnot efficiency of 36% and a power cost of 25 mills/kWh, the power cost for producing liquid hydrogen at its normal boiling point is $2.05/10^6 Btu ($0.28/kg).

No actual cost data exist for producing slush or solid hydrogen, since neither have been produced on a large-scale basis. To account for the additional complexity of such a plant, we recommend the use of data in Table 8 and Figure 7, assuming the percent of Carnot efficiency is two percentage points lower than the efficiency of a liquefier operating between the same temperature limits. Again, we estimate the O&M costs to be 5% of the installed capital costs. Cost estimates for slush and triple-point solid hydrogen, although speculative, are given by Parrish and Voth.[18]

In all of the above calculations it was assumed that the gas was normal hydrogen, i.e., 75% ortho- and 25% parahydrogen. It was also assumed that the liquid, slush, or solid was equilibrium hydrogen, which is nearly 100% parahydrogen. Based on the ideal work of liquefaction, roughly 15% of this work is due to the ortho-para conversion. Because the heat of conversion from ortho to para is exothermic,[19] failure to make the conversion will increase the boiloff rate of the liquid. Table 9 shows the potential reduction in the ideal work of liquefaction by not converting the hydrogen to the equilibrium concentration at 20 K. Figure 8 shows the fraction of liquid hydrogen boiled off as a function of storage time and initial ortho content; this is the boiloff due to ortho-para conversion only. These calculations indicate that it may be desirable to avoid conversion to equilibrium hydrogen if the liquid is to be stored for short periods of time.

TABLE 9

Reduction in Ideal Work of Liquefaction Due to Incomplete Conversion of Ortho- to Parahydrogen

Ortho content of liquid hydrogen, %	Reduction in ideal liquefaction energy,[a] %
0.211	0.0
5.0	2.4
10.0	4.2
20.0	7.1
30.0	9.5
40.0	11.4
50.0	13.0
60.0	14.3
75.0	15.4

[a] Ideal liquefaction energy as used herein includes the energy required to remove the heat of conversion (for conversion to 99.79% parahydrogen).

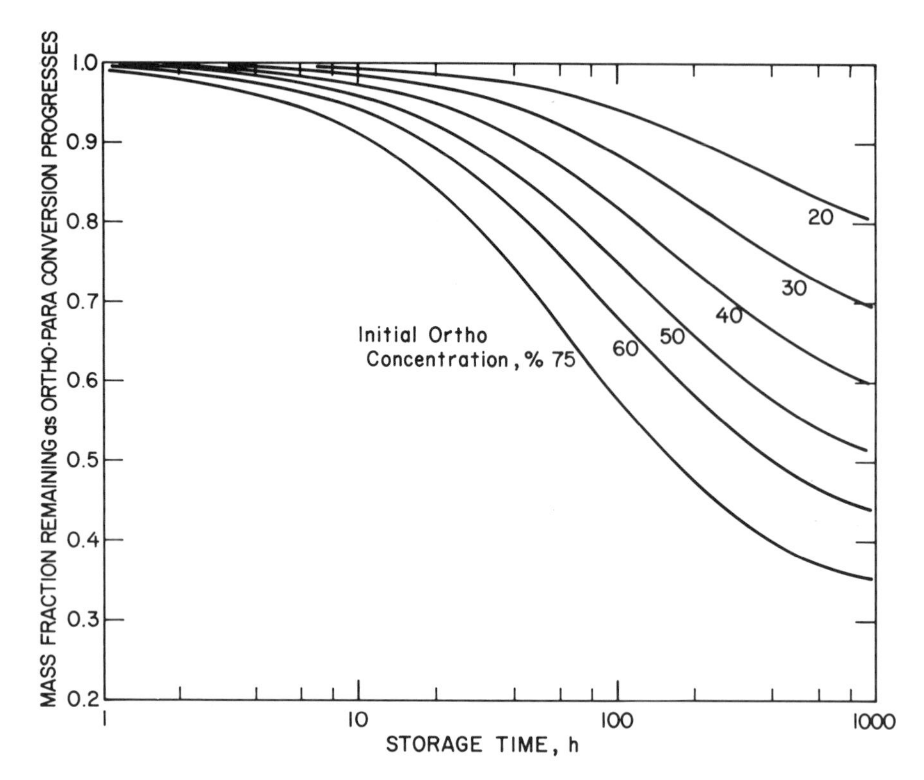

FIGURE 8. Fraction of the liquid hydrogen boiled off, due to ortho-para conversion, as a function of time.

1.3.3. Hydrogen Cost Credits

Both large-scale production and liquefaction offer possible cost credits which could lower the overall cost of hydrogen as a feedstock or fuel. If water is the source of hydrogen via electrolysis or thermochemical decomposition, large quantities of oxygen will become available. Figure 9 shows the reduction in the cost of hydrogen as a function of the by-product value of oxygen. The value of the oxygen could range anywhere from zero, due to oversupply or unavailable markets, to the current cost of oxygen from an air separation plant; the average cost of gaseous oxygen from air separation plants in 1974 was 0.76 ¢/lb (1.67 ¢/kg.)[20] Based on this cost, the by-product value of oxygen could lower the selling price of hydrogen as much as \$1.00/10^6 Btu (\$0.13/kg). However, to make by-product oxygen competitive with oxygen from an air separation plant, the source should be close to the consumer to minimize transmission costs. Beghi et al.[21] show that transmission costs for gaseous oxygen are double those of hydrogen for a specified volumetric flow rate.

The current demand for deuterium, the second most common isotope of hydrogen, is very low. However, the demand will increase significantly if the fusion power reactor becomes available, since the two most attractive fusion reactions use deuterium.[22] An attractive method of recovering deuterium, especially if liquid hydrogen is desired, is to use low-temperature distillation.[23] It is difficult to predict how much by-product credit could be given for deuterium; two estimates range from 13 to 26 ¢/10^6 Btu (1.75 to 3.5 ¢/kg).[24,25]

If liquid hydrogen is available but gaseous hydrogen at ambient temperature is desired, a fraction of the liquefaction energy can be recovered. The amount of liquefaction energy recovered depends upon the hydrogen application and system operating conditions. Thus, the fraction of liquefaction energy that can be recovered is system dependent. A recent study[26] shows that 5 to 50% of the actual work of liquefaction may be recovered; these energy-saving techniques would effectively lower the overall cost of liquid hydrogen.

1.4. HYDROGEN TRANSMISSION COSTS

The costs considered in this section are taken as independent of production and terminal costs.

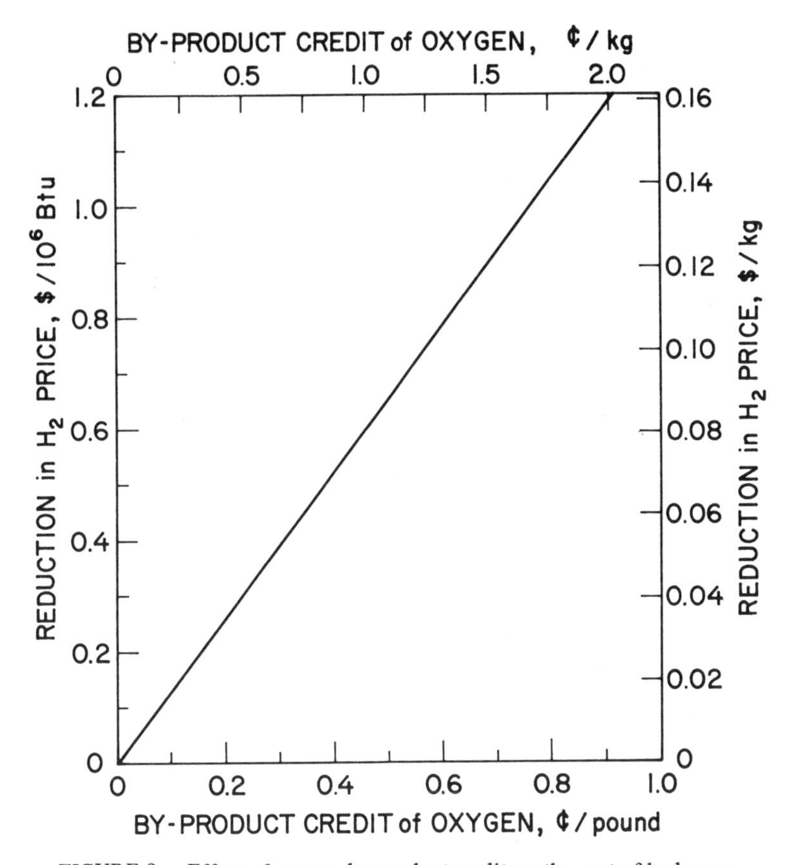

FIGURE 9. Effect of oxygen by-product credit on the cost of hydrogen.

Therefore, no allowance is made for the cost of delivering hydrogen to the transmission system in a suitable physical state (e.g., hydrogen gas at 2400 psia [16.55 MPa] and room temperature for transport by railroad tank car) or for the cost of receiving hydrogen into a suitable distribution network. In any system analysis, these terminal costs must be considered, and the supply-terminal costs are normally added to the production costs. The distribution costs in any fuel system are much more complex and are usually treated separately to account for the various modes of dispensing fuels. The transmission cost data given herein are based solely on the cost of moving hydrogen over relatively long distances, and are generally not applicable to piping distribution networks or short-haul freight deliveries. Distribution costs vary widely with each fuel application and are sometimes small enough to be absorbed by the operating budget of the receiving facility. In other cases, such as city gas pipe networks, the distribution costs are significant. Due to the scarcity of meaningful data, distribution costs are not included in this chapter, but a few appropriate references are cited for the convenience of the interested reader.

Unit transportation costs (e.g., $\$/10^6$ Btu-mi or \$/kg-km) are developed for each mode of hydrogen transport considered. *Actual* cost data are given wherever possible, and all cost *estimates* are based on the best available information.

1.4.1. Gas Pipeline

The estimated costs for piping hydrogen gas are taken from the extensive studies of Konopka and Wurm.[27] All estimates exclude the cost of the right-of-way and the initial gas compression costs, i.e., the cost of compressing the gas to line pressure at the entry to the line. These initial compression costs can be estimated from the data of Konopka and Wurm[27] (see also Beghi et al.[21] and Reynolds and Slager[28]) and from the compressor cost data given in Section 1.3.2. The cost of right-of-way varies with geographical locale and is much higher in urban areas; therefore, any attempt to estimate right-of-way costs would be meaningless. These costs must be assessed on an individual route basis; however, some right-of-way cost data for aboveground electrical power transmission are available from the Federal Power Commission.[29] Current figures[30] show that the cost of such a right-of-way varies from $220 to over $100,000/mi ($137 to over $62,137/km) depending upon the width of the right-of-way and its geographic location.

In the transmission model analyzed by Konopka and Wurm,[27] a transmission distance of 1000 U.S. statute mi (1609.34 km) was used, and compressor station spacing was varied from 50 to 300 mi (80.47 to 482.80 km). The transmission cost was not significantly affected by compressor spacing, and calculations were performed for 750 psia (5.17 MPa) and 2000 psia (13.79 MPa) pipelines. Optimized transmission costs were given for 50- and 100-mile (80.47- and 160.93-km) pipeline sections. These costs have been reduced to unit transmission costs and adjusted to reflect inflation, and are plotted on Figure 10. As would be expected, the transmission costs vary with pipeline pressure and pipeline size – for a given pipeline diameter, costs are lower and energy throughput is greater with the pipeline operating at the higher pressure. Leeth[57] has published cost comparisons for pipeline transmission of different fluids (including hydrogen gas).

While not directly applicable, some economic data for natural gas pipeline distribution costs are available and have been summarized by Gregory.[56] He estimates that pipeline distribution costs for hydrogen gas would be about 65% of the hydrogen transmission costs, and these distribution costs are about 25% greater than natural gas distribution costs (for equivalent energy throughput). Transmission and distribution costs for hydrogen are greater than those for natural gas because the HHV of hydrogen is about one third of the HHV of natural gas on a volumetric basis; therefore, hydrogen transmission requires higher pressure and/or larger pipelines and more compressor power than is required for natural gas (for equivalent energy throughput).

Offshore or undersea pipeline cost data are much more difficult to obtain. However, some oil and natural gas pipeline cost data are available and

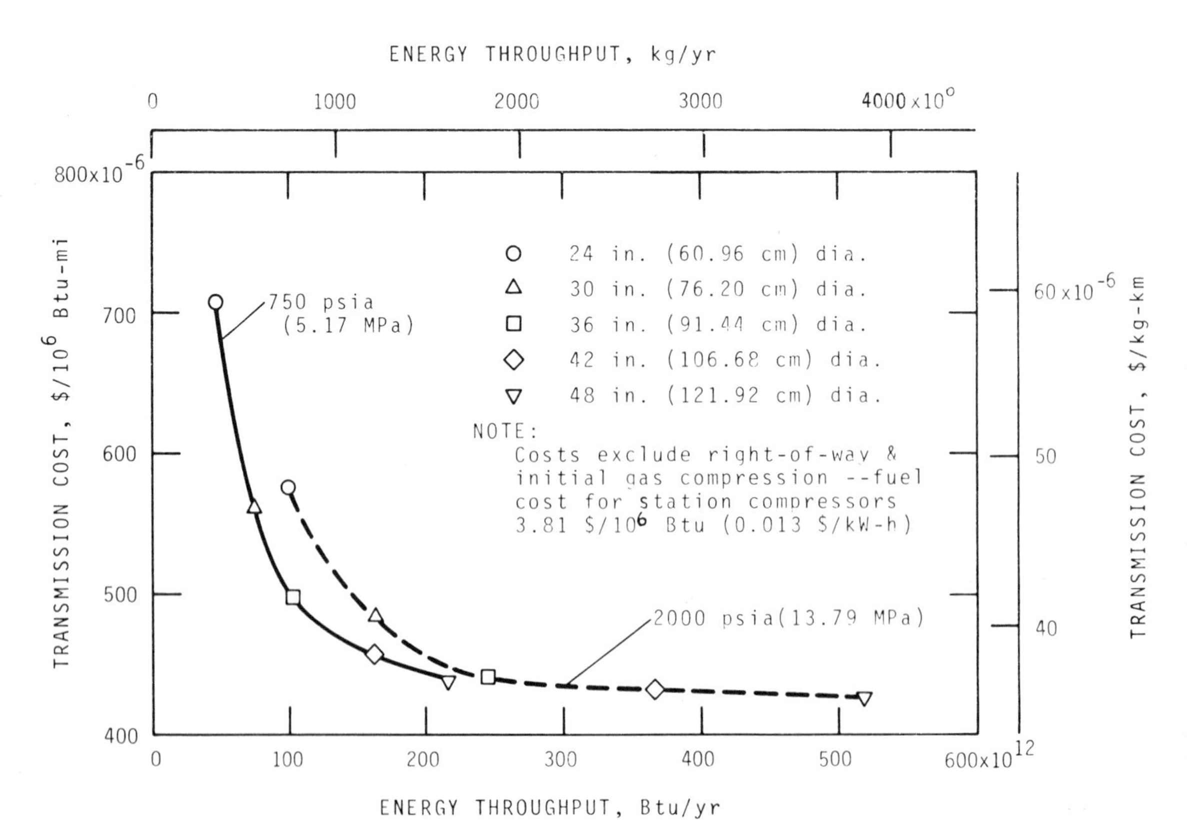

FIGURE 10. Pipeline transmission costs for hydrogen.

will be used to estimate the cost of laying underwater pipelines for transmission of hydrogen gas. Again, we will omit the initial gas compression costs, and in addition will assume that the lines are relatively short and that sea-based compressor stations are unnecessary. The latter assumption is perfectly valid for the available data. Hydrogen gas throughput capacities for each pipeline are taken from the calculations of Konopka and Wurm.[27]

Underwater pipeline installation costs vary widely with ocean-floor terrain, ocean temperature and depth, anchor protection, etc. There is a great deal of new activity in this field, e.g., the development of new equipment for offshore pipeline construction,[31,32] the use of epoxy-coated pipelines,[33] burial and protection studies,[34] and undersea towing[35] of prefabricated pipelines to offshore installation sites. The pipeline installation costs shown in Table 10 reflect considerable variation in the ease of installation. The most expensive installation occurred in Frigg Field of the North Sea (bottom line of Table 10). The cost of transmitting hydrogen through underwater pipelines was estimated as indicated in the footnotes of Table 10 and is tabulated therein. Comparison of these data with those given in Figure 10 (for the same throughput) indicates a much higher transmission cost for underwater pipelines.

1.4.2. Compressed Gas by Highway

Compressed hydrogen gas is currently transported over the highway by towing tube trailers with trucks. These tube trailers are fabricated by manifolding a number of large high-pressure steel tubes together and mounting them on a truck trailer. Operating pressures range from about 1750 to 2400 psia (12.07 to 16.55 MPa) with delivery capacities ranging from 38,000 to 115,000 SCF (90.1 to 272.6 kg). These trailers are constructed and operated according to existing Department of Transportation regulations,[38] and, in addition, must comply with municipal and state regulations for the transportation of flammable compressed gases.

Actual costs for transporting gaseous hydrogen by highway are reported herein: the costs vary with company accounting procedures; with the frequency, range, and capacity of delivery, and, to a lesser extent, with the geographic locale. The unit transportation costs given in Table 11 are obtained by summing the truck and tube trailer component costs and dividing by the quantity of delivered hydrogen. The transportation costs for government-owned equipment are considered upper limits, as these rates are full costs and apply to occasional deliveries. Similarly, the lease equipment rates can be considered lower limits for intermittent deliveries, as they do not include some potential charges (see footnotes in Table 11). Negotiated contracts for frequent or continuous deliveries would undoubtedly result in lower transportation costs.

The highest unit transportation costs accompany the smallest tube trailers (see top line of Table 11). Very few of these small tubes trailers are still used in over-the-road service. Some have been replaced by the larger tube trailers (115,000 SCF) which currently cost about $70,000 and are usually amortized in 20 years. These "jumbo" tube trailers weigh about 57,000 lb (25,855 kg). are constructed of high-strength steel (115,000 psi [793 MPs] tensile) and comply with Department of Transportation (DOT) specifications 3A, 3AX, 3AA, or 3AAX.[38] Thus, the as-fabricated cost is about $70,000/57,000 = $1.23/lb of steel ($2.71/lb of steel). Department of Transportation regulations[38] require periodic inspections of safety relief devices and periodic hydrostatic pressure tests of hydrogen-gas containers: therefore, the budgeted maintenance costs for these trailers must include such inspection costs.

As indicated in Table 11, the current cost of transporting compressed hydrogen gas over the highway by using large tube trailers ranges from $0.041 to $0.068/10^6 Btu-mi ($0.0034 to $0.0057/kg-km).

1.4.3. Compressed Gas by Railway

Compressed hydrogen gas is not currently being shipped by rail, although suitable railroad tank cars and Department of Transportation regulations[38] exist. These tank cars are built much like the highway tube trailers, i.e., a number of seamless steel cylinders are manifolded and mounted on a special railroad car frame. The tank cars must be fabricated according to Department of Transportation 107A specifications[39] and must be equipped with Department-approved ignition devices that will instantly ignite any hydrogen that may be discharged through an approved safety relief device. A number of railroad tank cars, of nearly identical construction, are currently being used to ship compressed helium gas. The tank car

TABLE 10

Cost of Transmitting Hydrogen Gas Through Underwater Pipelines

	Length		Diameter		Installed pipeline cost (estimated)		H_2 energy throughput[a]		Estimated unit transmission cost[b]		
Pipeline location	mi	km	in.	cm	\$/mi	\$/km	Btu/year	kg/year	$\$/10^6$ Btu-mi	\$/kg-km	Reference
Offshore Texas	28.9	46.5	36	91.44	1.08×10^6	0.67×10^6	101×10^{12}	0.75×10^9	1.93×10^{-3}	161×10^{-6}	36
Offshore Texas	21.9	35.2	24	60.96	0.64	0.40	45	0.33	2.61	218	36
Offshore Texas/ Louisiana	45.8	73.7	30	76.20	0.96	0.60	72	0.54	2.39	200	36
Offshore Texas	71.4	114.9	36	91.44	0.96	0.60	101	0.75	1.72	144	36
Offshore Texas/ Louisiana	18.7	30.1	30	76.20	0.30	0.19	42*	0.31	1.32	110	36
Offshore Texas	176.0	283.2	~30[c]	~76.20[c]	1.23	0.76	72	0.54	3.03	253	37
Offshore Texas/ Louisiana	200.0	321.9	~36[c]	~91.44[c]	1.72	1.07	101	0.75	3.07	257	37
Offshore Scotland (Frigg Field)	200.0	321.9	34	86.36	5.00	3.11	90	0.67	10.00	836	37

[a]Estimated from the data of Konopka and Wurm[27] using 750 psia pipeline operating pressure. (*~375 psia pipeline)
[b](Installed pipeline cost) × ACR ÷ throughput: where ACR = 0.18 = FCR (0.13) + O & M (0.03) + IRRI (0.02); FCR = CRF (0.11) + T & I (0.02); i = 0.09, n = 20-year life.
[c]Average pipeline diameter.

TABLE 11

Cost of Transporting Hydrogen Gas Over the Highway by Truck Tube Trailer

Mode of operation	Truck cost		Tube trailer cost		Delivery capacity		Delivery distance		Delivery pressure		Unit transportation cost	
	$/mi	$/km	$/mi	$/km	SCF	kg	mi	km	PSIA	MPa	$\$/10^6$ Btu-mi	$/kg-km
Government-owned	1.40–2.04	0.87–1.27	0.26	0.16	38,000	90.1	300–1100	483–1770	1825	12.58	0.137–0.190	0.0114–0.0159
Government-owned	1.40–2.04	0.87–1.27	0.46	0.29	115,000	272.6	300–1100	483–1770	1900–2400	13.10–16.55	0.051–0.068	0.0043–0.0057
Lease-rental	1.10[a]	0.68[a]	0.38[b]	0.24[b]	82,000	194.3	>300	>483	1750	12.07	0.057[a]	0.0047[a]
Lease-rental	1.10[a]	0.68[a]	0.40[a]	0.25[a]	115,000	272.6	>300	>483	1900–2400	13.10–16.55	0.041[a]	0.0034[a]
Industry-owned	1.04–1.25[a]	0.65–0.78[a]	0.46[b]	0.29[b]	115,000	272.6	>50	>80	1900–2400	13.10–16.55	0.041–0.047[a]	0.0034–0.0039[a]

Note: All rates in this table are based on intermittent delivery and one-way mileage.

[a]Costs do not include driver layovers or demurrage (also, the lease rates do not include round trip mileage from the lease center to one of the destination points).
[b]Government owned.

costs and operating costs should be almost identical for helium and hydrogen shipments; therefore, we will use the helium shipment cost data to estimate the cost of shipping hydrogen gas by railroad tank car.

A railroad tank car weighs about 220,000 lb (99,792 kg), and replacement cost is estimated at approximately \$200,000 – \$20,000 for the car running gear and \$180,000 for the tube and cabinet assembly. Then, the current as-fabricated cost is estimated at 200,000/220,000 = \$0.91/lb of steel (\$2.00/kg of steel). Each tank car has a nominal hydrogen storage capacity of 210,000 SCF (497.7 kg) at operating pressures of 2450 psia (16.89 MPa). The Department of Transportation permits much higher operating pressures (1.57 X) and fill densities when these tank cars are used in helium gas service. Thus, the tank car discussed above would be permitted to transport about 300,000 SCF (1421 kg) of gaseous helium at approximately 3850 psia (26.54 MPa).

Table 12 summarizes the estimated costs for transporting gaseous hydrogen by railroad tank car. All of the cost data are extrapolated from actual costs incurred in moving compressed helium gas by tank car. Most of these data were supplied by the Bureau of Mines,[39] and the costs tabulated are for government-owned tank cars that are leased to the user. Lease rates are approximately \$0.21/round-trip mile (\$0.13/round-trip km) for helium service, and idle car (at delivery site) charges are \$25/day. In the cost estimates for hydrogen service, the helium service lease rates were increased by 10% to account for increased cost of car construction and periodic inspections. The idle car charges were ignored, as they are relatively insignificant with respect to freight and ownership charges. Also, very little idle time is experienced or anticipated with tank car deliveries. Freight rates from the mid-U.S. to the West Coast are higher than from the mid-U.S. to the East Coast, as indicated in Table 12.

The unit transportation costs in Table 12 are obtained by adding the freight and tank car ownership costs and dividing by the quantity of delivered hydrogen product.

1.4.4. Compressed Gas by Ship or Barge

Compressed hydrogen gas is not commonly shipped overseas in quantity; however, bulk shipments of compressed helium gas and liquid helium are relatively common. As with railway transport, the helium shipments provide the economic framework that permits us to estimate the cost of marine transportation of hydrogen gas. Gaseous helium has been transported from Seattle, Washington to Whittier, Alaska by loading railroad tank cars on ocean-going barges. Compressed helium gas has also been transported from New York to Europe by loading highway tube trailers on cargo ships. Highway tube trailers and railroad tank cars are transported on the ship's deck.

The most common means of shipping gaseous helium overseas is by the use of specially "packaged" container units. These packaged units are 8 ft X 8 ft X 20 ft (2.44 m X 2.44 m X 6.10 m) or 8 ft X 8 ft X 40 ft (2.44 m X 2.44 m X 12.19 m), can be stacked six high, and may be transported in the ship's cargo hold. Hydrogen shipments in such container units are not currently permitted[40] below deck on cargo ships. The nominal helium storage capacity of these container units is 120,000 SCF (568.3 kg) and 240,000 SCF (1136.5 kg) for the smaller and larger units, respectively. These units are fabricated according to Department of Transportation 107A specifications[38] and must operate at lower pressures if filled with hydrogen gas; therefore, the helium operating pressure of 2640 psia (18.20 MPa) is reduced to 2640/1.57 = 1682 psia (11.60 MPa) for hydrogen service. The corresponding hydrogen storage capacities are 76,433 SCF (181.1 kg) and 152,866 SCF (362.3 kg) for the smaller and larger container units, respectively.

The cost estimates for barging and shipping compressed hydrogen gas are listed in Table 13. All transportation costs are based on single-unit cargo rates and would undoubtedly be lowered by volume shipping and/or negotiation with the carriers. Current DOT regulations[40] permit on-deck shipment of compressed hydrogen gas in approved containers on cargo vessels and in approved railway tank cars on trainships.

1.4.5. Liquid Pipeline

Only one cross-country liquid hydrogen pipeline has been installed in the U.S., and replacement cost is somewhat uncertain. This vacuum-jacketed pipeline is about 1500 ft (0.46 km) long with a nominal 6-in (15.24 cm) diameter inner line and is installed at Kennedy Space Center in Florida. It is felt that the best cost estimates of long-distance vacuum-insulated pipelines were presented in a study[41] on superconducting power transmission

TABLE 12

Cost of Transporting Hydrogen Gas by Railroad Tank Car

Railroad tank car route	Freight charges[a]		Tank car cost[b]		Delivery capacity		Delivery distance		Delivery pressure		Delivery time (one-way) in days	Unit transporation cost	
	\$/mi	\$/km	\$/mi	\$/km	SCF	kg	mi	km	PSIA	MPa		\$/10⁶ Btu-mi	\$/kg-km
Keyes, Okla. to Kennedy Space Center, Fla. (and return)	1.46	0.91	0.46	0.29	210,000	497.7	1725	2776	2450	16.89	10–14	0.029	0.0024
Keyes, Okla. to Chatsworth, Cal. (and return)	1.99	1.24	0.46	0.29	210,000	497.7	1385	2229	2450	16.89	7–8	0.037	0.0031
Keyes, Okla. to Seattle, Wash. (and return)	1.92	1.19	0.46	0.29	210,000	497.7	1700	2736	2450	16.89	8–10	0.036	0.0030
Keyes, Okla. to Camden, N.J. (and return)	1.49	0.93	0.46	0.29	210,000	497.7	1728	2781	2450	16.89	10–12	0.029	0.0024

Note: All rates in this table are based on intermittent delivery of single cars and one-way mileage.

[a] Up-to-date railway freight rates are available from the carriers and are regulated by the Department of Transportation. The freight charges are based on a minimum delivered product weight of 40,000 lb (18,144 kg). Reduced rates can be negotiated for train units (multicar shipments) within 90 days.
[b] Costs shown are lease rates for government-owned tank cars.

TABLE 13

Cost of Transporting Hydrogen Gas by Barge or Ship

Container description and ship route	Freight charges		Container cost		Delivery capacity		Delivery distance		Delivery pressure		Unit transportation cost	
	$/mi	$/km	$/mi	$/km	SCF	kg	mi	km	PSIA	MPa	$/10⁶ Btu-mi	$/kg-km
Railroad tank car: Seattle, Wash. to Whittier, Alaska by barge (and return)	1.89	1.17	0.46[a]	0.29[a]	210,000	497.7	1500	2414	2450	16.89	0.035	0.0029
Highway tube trailer: New York to Southampton, England by cargo ship (and return)	1.93	1.20	0.46[a]	0.29[a]	115,000	272.6	3630	5842	2100	14.48	0.065	0.0055
Container unit:[b] New York to Southampton, England by cargo ship (and return)	0.87	0.54	0.29[c]	0.18[c]	76,500	181.3	3630	5842	1682	11.60	0.048	0.0040
Container unit:[b] New York to Southampton, England by cargo ship (and return)	1.62	1.01	0.54[c]	0.34[c]	153,000	362.6	3630	5842	1682	11.60	0.045	0.0037
Container unit:[b] New York to Antwerp, Belgium by cargo ship (and return)	1.65	1.03	0.54[c]	0.34[c]	153,000	362.6	3880	6244	1682	11.60	0.045	0.0038

Note: All rates in this table are based on single container units, tube trailers, or tank cars and one-way mileage.

[a]Assumes lease rates for government-owned tank cars and tube trailers applicable to land or water mileage.
[b]Manifolded and packaged tube bundles.
[c]Estimated ownership costs for industry-owned containers.

lines. Power transmission lines are not constructed as ordinary liquid transfer pipes; therefore, it was necessary to extract component labor and materials cost estimates from this study[41] to combine with current prices to obtain installed-pipeline cost estimates.

Table 14 summarizes the liquid pipeline component cost estimates. The cost of prefabricated superinsulated piping was estimated from cost data supplied by commercial firms and is based on bulk manufacture of more than 100 pieces of 60-ft (18.29-m) lengths. The trenching costs are based on a suburban installation[41] and include 69 street crossings, two parkway borings, and two stream crossings. The refrigeration requirements were taken from optimized calculations,[42] and the refrigeration costs were updated from the publication by Strobridge.[15] Pump costs were taken from current prices of similar pumps used to transfer liquefied natural gas. Maintenance rates were assumed to be identical with those of the utility and process industries (≈0.03I), and the cost of pump and refrigeration energy was taken at $0.025/kWh. The annual charge rate (ACR) was taken at 18% (see footnotes of Table 10).

Using data from Table 14 and other calculations,[42] the effects of pipeline size on unit transportation costs were estimated. These results are given in Table 15 and plotted on Figure 11.

1.4.6. Liquid by Highway

In the U.S., liquid hydrogen is most commonly transported by well-insulated truck trailers. These trailers are authorized for highway service under special permit by the Department of Transportation[38] and range in capacity from 8000 to 13,000 gal (30.28 to 49.21 m^3). Under normal conditions, these truck trailers will experience a pressure rise of less than 1 atm within a delivery range of 3000 mi (4828 km). The cost data summarized in Table 16 are based upon current delivery charges.

Most insulated truck trailers now deliver around 12,300 gal (46.56 m^3) or 7200 lb (3265.9 kg) of liquid hydrogen. These trailers may be purchased

TABLE 14

Cost Components of a Liquid Hydrogen Pipeline

Description or item	Cost
Materials: Prefabricated vacuum-insulated pipeline, 60-ft (18.29-m) lengths, complete with end fixtures and expansion relief ($25/in of inner pipe diameter per foot of length [$32.30/cm-m])	$18.45 × 10⁶
Field labor: fitting, welding, leak checking (34,450 man-hr @ $15/man-hr)	0.52
Trenching: trench and backfill (0.95), street crossings (0.18), parkway borings (0.08), stream crossings (0.39), position pipes (0.44), valve boxes (0.80), evacuate and cold test (0.11), alarms and corrosion resistance (0.09)	3.04
Refrigeration: 35% efficient, 9.32-mi (15.0-km) spacing, installed power (P) = 1290 kW (heat leak) + 18,800 kW (pump energy) = 20,090 kW; (cost = 7560 $P^{0.7}$ × 2 refrigerators)	15.55
Pump: 60% efficient, 9.32-mi (15.0-km) spacing, 32 lb/sec (14.52 kg/sec) flow rate @ 13 atm head rise, integral pump-motor ($125/horsepower × 600 horsepower × 2 pumps)	0.15
Pump and refrigeration station house: high-bay corrugated metal building with overhead cranes, explosion-proof electrical equipment and alarm systems (20 m × 20 m × 10 m high @ $500/m² floor space × 2 houses)	0.40
Total capital investment	$38.11 × 10⁶
Maintenance: (@ 3% per year), (0.03 × $38.11 × 10⁶)	1.14 × 10⁶ $/year
Energy: 40,180 kW (refrigeration) + 876 kW (pumps) = 41,056 kW for 328 days/year @ 24 hr/day (323.19 × 10⁶ kWh × 0.025 $/kWh)	8.08
FCR (0.13) + IRRI (0.02); (0.15 × $38.11 × 10⁶)	5.72
Total annual charges	$14.94 × 10⁶ $/year
Unit transportation cost: (total annual charges ÷ throughput) ÷ pipe length (14.94 × 10⁶ $/year ÷ 55.37 × 10¹² Btu/year ÷ 18.64 mi)	0.0145 $/10⁶ Btu-mi 0.00121 $/kg-km

Note: Right-of-way costs excluded. Superinsulated pipeline is 18.64 mi (30.0 km) long; heat leak is 2.08 Btu/hr-ft (2 W/m); requires a field joint every 60 ft (18.29 m); requires a refrigerator and a liquid pump every 9.32 mi (15.0 km); liquid enters pipe at 14 atm and 32.9 R (18.3 K) and exits at 1 atm and 36 R (20 K); cold pipe is 7.50 in (19.05 cm) inside diameter; throughput is 55.37 × 10¹² Btu/year (411.54 × 10⁶ kg/year); 328 day/year operation.

TABLE 15

Cost of Transmitting Liquid Hydrogen Through Underground Vacuum-insulated Pipelines

Description or item	4-in. (10.16-cm) I.D. pipe		7.5-in. (19.05-cm) I.D. pipe		11.22-in. (28.50-cm) I.D. pipe	
Heat leak	11.11 Btu/hr-ft (1.07 W/m)		2.08 Btu/hr-ft (2 W/m)		3.12 Btu/hr-ft (3 W/m)	
Pump and refrigerator spacing	6.21 mi (10 km)		9.32 mi (15 km)		12.43 mi (20 km)	
Liquid inlet temperature	32.67 R (18.15 K)		32.89 R (18.27 K)		32.98 R (18.32 K)	
Liquid exit temperature	36 R (20 K)		36 R (20 K)		36 R (20 K)	
Liquid inlet pressure	14 atm		14 atm		14 atm	
Liquid exit pressure	1 atm		1 atm		1 atm	
Mass flow rate	7.65 lb/sec (3.47 kg/sec)		32.01 lb/sec (14.52 kg/sec)		79.37 lb/sec (36.00 kg/sec)	
Annual throughput	13.23×10^{12} Btu/year (98.34×10^6 kg/year)		55.36×10^{12} Btu/year (411.49×10^6 kg/year)		137.27×10^{12} Btu/year (1020.21×10^6 kg/year)	
			Cost[a]			
	$/in.-ft	$/cm-m	$/in.-ft	$/cm-m	$/in.-ft	$/cm-m
Materials	25.00	32.29	25.00	32.29	25.00	32.29
Field labor	0.81	1.05	0.70	0.90	0.64	0.83
Trenching	6.62	8.55	4.10	5.30	3.12	4.03
Refrigeration	22.20	28.68	21.06	27.20	19.77	25.53
Pump	0.13	0.17	0.20	0.26	0.25	0.32
Pump house	0.46	0.59	0.54	0.70	0.58	0.75
Total capital investment	55.22	71.33	51.60	66.65	49.36	63.75
	$/in.-ft-year	$/cm-m-year	$/in.-ft-year	$/cm-m-year	$/in.-ft-year	$/cm-m-year
Maintenance	1.66	2.14	1.55	2.00	1.48	1.91
Energy	7.57	9.78	10.95	14.14	13.44	17.36
FCR + IRRI	8.28	10.70	7.74	10.00	7.40	9.56
Total annual charges	17.51	22.62	20.24	26.14	22.32	28.83
	$/10^6 Btu-mi	$/kg-km	$/10^6 Btu-mi	$/kg-km	$/10^6 Btu-mi	$/kg-km
Unit transportation cost	0.0280	2.34×10^{-3}	0.0145	1.21×10^{-3}	0.0096	0.80×10^{-3}

[a]Right-of-way costs excluded, 328 day/year operation, costs expressed in dollars per inch of inside diameter of cold pipe per foot of length.

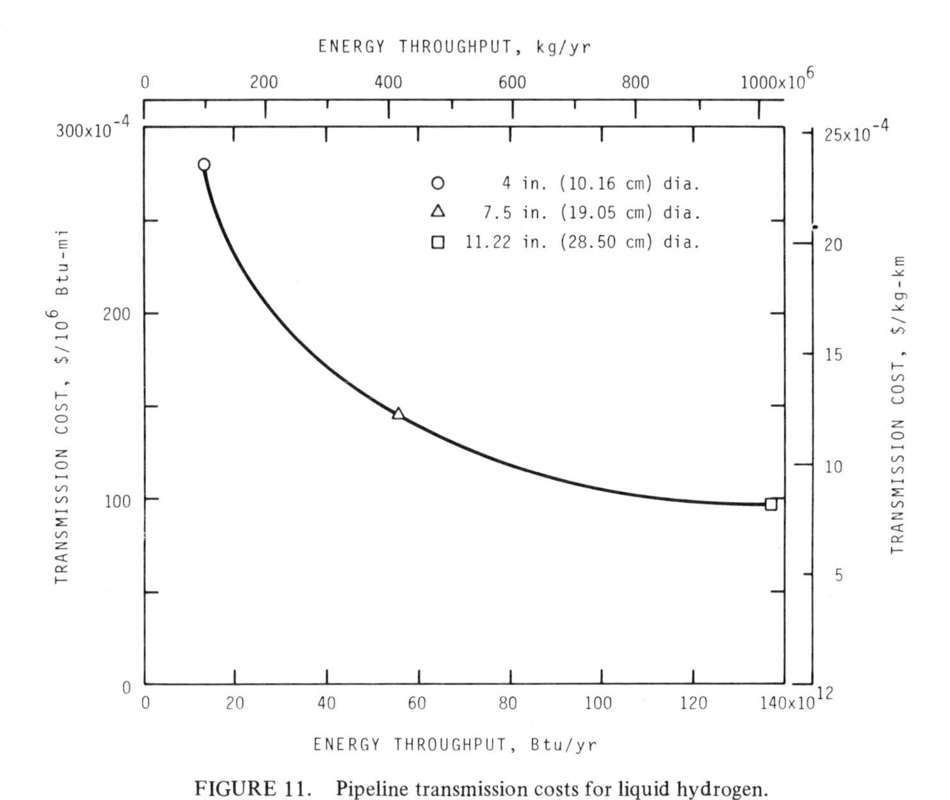

FIGURE 11. Pipeline transmission costs for liquid hydrogen.

TABLE 16

Cost of Transporting Liquid Hydrogen Over the Highway by Insulated Tank Truck Trailers

	Truck cost		Tank trailer cost		Delivery capacity		Delivery distance		Unit transportation cost	
Mode of operation	\$/mi	\$/km	\$/mi	\$/km	lb	kg	mi	km	\$/10⁶ Btu-mi	\$/kg-km
Government-owned[a]	1.40	0.87	1.88	1.17	7200	3266	1100	1770	0.0075	0.00063
Lease-rental[a]	1.10[c]	0.68[c]	1.90[c]	1.18[c]	7200	3266	>250	>402	0.0068[c]	0.00057[c]
Industry-owned[a]	1.25[c]	0.78[c]	2.00[c]	1.24[c]	7200	3266	1200	1931	0.0074[c]	0.00062[c]
Industry-owned[b]	1.25[c]	0.78[c]	1.35[c]	0.84[c]	7200	3266	600	966	0.0059[c]	0.00049[c]

[a]Rates based on intermittent delivery and one-way mileage.
[b]Rates based on frequent delivery between two cities and one-way mileage.
[c]Costs do not include driver layovers or demurrage (also, the lease rates do not include round-trip mileage from the lease centers to one of the destination points).

for \$125,000 to \$150,000, and amortization periods range from 7 to 11 years. Maintenance costs on these vehicles are usually high, and annual charges may easily exceed 25% of the initial investment. For example, using a 10-year life and 9% interest, we compute a CRF $\approx$ 0.16. Adding 3% for T&I, 6% for O&M, and 2% for IRRI, we obtain an ACR of 27% of the capital investment. Thus, once again, the unit transportation cost of hydrogen is highly dependent upon delivery distance and frequency. The economic advantages of frequent deliveries are made apparent by comparing the cost data in the two bottom lines of Table 16.

1.4.7. Liquid by Railway

Liquid hydrogen may be transported in well-insulated railroad tank cars under current Department of Transportation regulations.[38] These tank cars have nominal capacities of 28,300 gal (107.13 m^3), are pressure-relieved at 17 psig (0.117 MPa), and under normal conditions do not vent within a delivery range of 3000 mi (4828 km). Vented hydrogen is mixed with excess air to form mixtures well below the lower flammable limit.

Actual cost data for shipment in these cars are scarce, and only one U.S. industrial firm is currently shipping liquid hydrogen by railroad car. This company considers its cost records proprietary; consequently, we are compelled to estimate the cost of railway transportation by using other information available to the public. Our cost estimates are prepared by using 1) actual railroad tank car (gaseous helium) delivery times, 2) estimated tank car replacement and operating costs, and 3) a simplistic industrial accounting method. Also, European cost estimates[43] for transport of liquefied natural gas (LNG) by railroad are adjusted to indicate the cost of moving liquid hydrogen by railway. These updated European cost estimates compare favorably with those derived in our independent analysis. The comparative cost data are summarized in Table 17.

The initial cost of a 28,300-gal (107.13-m^3) superinsulated railroad tank car is estimated at \$270,000 – \$250,000 for the tank and \$20,000 for the car running gear. A 9-year lifetime is assumed at an interest rate of 9% to produce a CRF $\approx$ 0.17. Adding T&I (0.02), O&M (0.05), and IRRI (0.02), we obtain an ACR of 26% of the capital investment. Assuming that the tank car is in service 300 days/year, the cost of ownership is (\$270,000 × 0.26) ÷ 300 = \$234/day. Thus, it is apparent that delivery time will have a pronounced effect on transportation charges.

Freight rates in the U.S. are highly dependent upon mileage, loaded-car direction (East or West), number and weight of loaded cars, and regularity of shipments. The freight charges listed in Table 17 are based on freight rates published in current railway freight tariffs. These tariffs apply to 40,000-lb (18,144-kg) minimum product delivery and are given as follows: \$0.0623/lb (\$0.1373/kg) from Los Angeles to Chicago and return; \$0.0959/pound (\$0.2114/kg) from Denver to Kennedy Space Center and return. Lower rates for large-volume shipments may be negotiated with the carriers. We estimate that special trains (25 cars or more) could halve the delivery times with a 20% reduction in freight rates. The foregoing input data are reflected in the cost figures listed in Table 17.

The German railway cost estimates for transporting LNG were kept essentially intact; however, these costs were modified to account for 300 day/year service, higher costs (15%) of liquid hydrogen tank-car fabrication, and inflation. All other features of the German railway transportation system were used as proposed by Backhaus and Janssen.[43] Briefly, their vacuum-insulated tank cars deliver 15,850 gal (60 m^3) of product with an updated ownership cost of \$85.3/day. The current replacement cost of these tank cars is estimated at \$167,000 and they were amortized[43] over 20 years at 12% interest. The German freight charges listed in Table 17 have also been adjusted to account for inflation. The special train units were allowed[43] a 25% reduction in freight rates and much shorter delivery times.

1.4.8. Liquid by Ship or Barge

Minimal information exists relative to the marine transportation of liquid hydrogen. NASA has, under special permit of the U.S. Coast Guard, barged liquid hydrogen over distances of about 52 mi (83.7 km) offshore Louisiana-Mississippi. The barge was NASA-owned and was fabricated by equipping a surplus U.S. Navy hull with a vacuum-insulated vessel. Consequently, replacement costs are somewhat uncertain, and actual operating costs are not well defined, due to the infrequent operation of these barges and government ownership. No other cost data for marine transportation of liquid hydrogen are available.

Once again, as with railroad tank cars, we will resort to the use of available LNG cost data to estimate the cost of transporting liquid hydrogen by ship or barge. Cost data are also deduced from the information available on the NASA barging operation. Legal provisions for the marine transportation of liquid hydrogen are set forth in Department of Transportation regulations.[40] Special permission of the U.S. Coast Guard Commandant is required. Cost data were derived from various sources and adjusted to account for inflation and the higher (10%) cost of liquid hydrogen vessel fabrication. The resultant cost estimates are summarized in Table 18, and unit transportation costs are plotted against the product of delivery capacity and distance on Figure 12.

Marine transportation costs are strongly influ-

TABLE 17

Cost of Transporting Liquid Hydrogen by Insulated Railroad Tank Car

Railroad tank car route	Freight charges[a]		Tank car cost[b]		Delivery capacity		Delivery distance		Frequency of delivery (days)	Unit transportation cost	
	$/mi	$/km	$/mi	$/km	lb	kg	mi	km		$/10⁶ Btu-mi	$/kg-km
Los Angeles, Cal. to Chicago, Ill. (and return); single car	1.19	0.74	2.90	1.80	14,900	6,759	2100	3380	26	0.0045	0.00038
Los Angeles, Cal. to Chicago, Ill. (and return); 25-car train	23.73	14.75	36.21	22.50	372,500	168,963	2100	3380	13	0.0026	0.00022
Denver, Colo. to Kennedy Space Center, Fla. (and return); single car	2.11	1.31	4.50	2.80	14,900	6,759	1821	2931	35	0.0073	0.00061
Denver, Colo. to Kennedy Space Center, Fla. (and return); 25-car train	42.13	26.18	57.83	35.93	372,500	168,963	1821	2931	18	0.0044	0.00037
Wilhelmshaven, Germany to Cologne (and return); single car[c]	2.17	1.35	1.54	0.96	9,300	4,218	222	357	4	0.0065	0.00054
Wilhelmshaven, Germany to Cologne (and return); 25 car train[c]	40.65	25.26	19.22	11.94	232,500	105,460	222	357	2	0.0042	0.00035
Wilhelmshaven, Germany to Basel, Switzerland (and return); single car[c]	1.50	0.93	1.16	0.72	9,300	4,218	515	829	7	0.0047	0.00039

TABLE 17 (continued)

Cost of Transporting Liquid Hydrogen by Insulated Railroad Tank Car

Railroad tank car route	Freight charges[a]		Tank car cost[b]		Delivery capacity		Delivery distance		Frequency of delivery (days	Unit transportation cost	
	$/mi	$/km	$/mi	$/km	lb	kg	mi	km		$/10^6 Btu-mi	$/kg-km
Wilhelmshaven, Germany to Basel, Switzerland (and return); 25-car train[c]	27.92	17.35	12.42	7.72	232,500	105,460	515	829	3	0.0028	0.00023

Note: All rates in this table are based on one-way mileage.

[a]Up-to-date railway freight rates are available from the carriers and are regulated by the federal governments of the U.S. and Germany. In the U.S., the freight charges are based on a minimum delivered product weight of 40,000 lb (18,144 kg). Reduced rates can be negotiated for train units (multicar shipments) within 90 days.
[b]Costs shown are based on estimated fabrication and operating costs.
[c]Source data from Backhaus and Janssen.[43]

TABLE 18

Cost of Transporting Liquid Hydrogen by Barge or Ship

Barge or ship route	Freight charges		Barge or ship cost		Delivery capacity		Delivery distance		Frequency of delivery (days)	Unit transportation cost		Reference
	\$/mi	\$/km	\$/mi	\$/km	lb	kg	mi	km		$\$/10^6$ Btu-mi	\$/kg-km	
River barge: Michoud, La. to Bay St. Louis, Miss. (and return)[a]	46.15	28.68	52.62	32.70	284,000	128,820	52	83.7	1	0.0057	0.00048	44
Ocean barge: Everett, Mass. to Brooklyn, N.Y. (and return)[b]	41.18	25.59	75.29	46.78	665,000	301,639	255	410.4	3	0.0029	0.00024	46, 47
Self-propelled river tanker: Rhine River; route not specified[c]	19.03	11.82	43.46	27.00	371,000	168,283	124	200.0	2	0.0028	0.00023	43
Self-propelled river tanker: Rhine River; route not specified[c]	14.24	8.85	32.53	20.21	371,000	168,283	497	800.0	6	0.0021	0.00018	43
River barge: Rhine River; route not specified[c]	31.20	19.39	121.29	75.37	1,088,000	493,509	124	200.0	2	0.0023	0.00019	43
River barge: Rhine River; route not specified[c]	23.36	14.52	90.78	56.41	1,088,000	493,509	497	800.0	6	0.0017	0.00014	43
Ship: route not specified[c]	32.14	19.97	367.95	228.63	11×10^6	5×10^6	2,300	3,701	12	0.00060	0.000050	48
Ship: route not specified[c]	32.74	20.34	543.30	337.59	19.4×10^6	8.80×10^6	2,300	3,701	11	0.00049	0.000041	48
Ship: Arabian Gulf to the U.S. East Coast (and return)[c]	55.09	34.23	380.20	236.25	19.4×10^6	8.80×10^6	14,950	24,060	58	0.00037	0.000031	49

Note: All rates exclude terminal costs and are based on one-way mileage (U.S. statute).

[a]Cost data based on liquid hydrogen barging experience.
[b]Cost data based on LNG barging experience.
[c]Cost data extrapolated by others from LNG shipping experience.

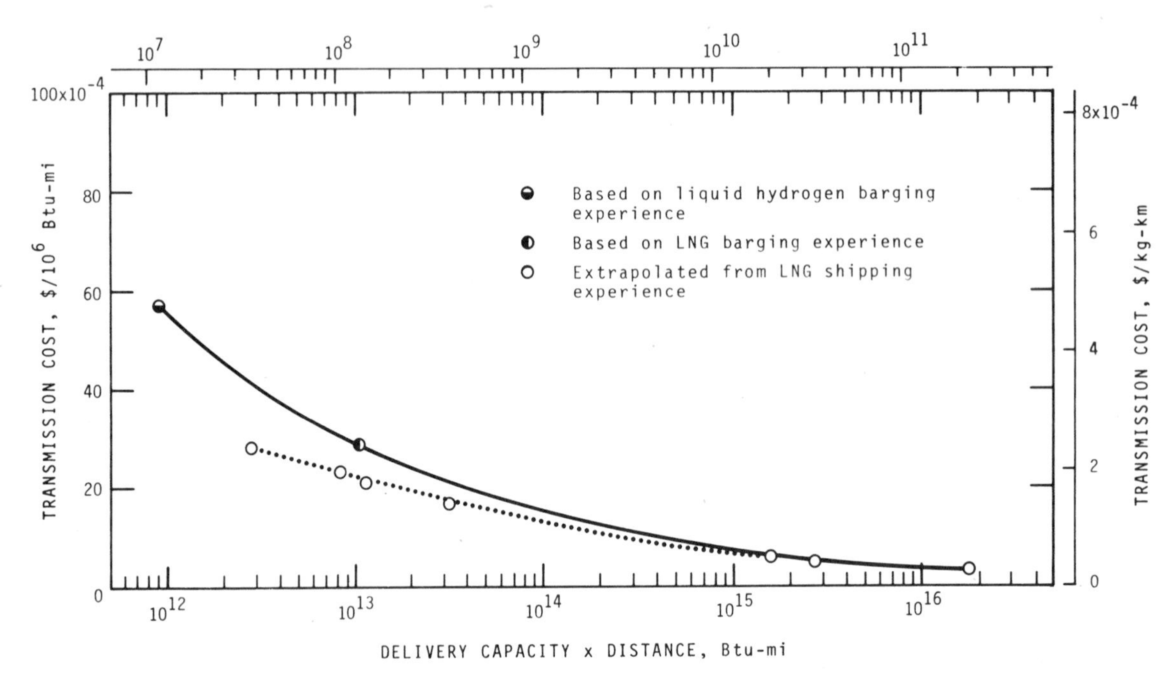

FIGURE 12. Marine transportation costs for liquid hydrogen – effects of delivery capacity and distance.

enced by delivery capacity and distance, as shown by the results in Table 18 and Figure 12. All of the open-circle data points plotted on Figure 12 represent cost extrapolations from intermediate-size (11 × 10^6 lb [5 × 10^6 kg] liquid hydrogen equivalent capacity) LNG ships. The estimates for larger capacities and distances seem reasonable, but the extrapolations to smaller capacities and distances are questionable. The solid-curve costs on Figure 12 may well be reduced by reducing the freight rates (see Table 18) on a *busy* barge; however, it is unlikely that the low dotted-curve costs on Figure 12 could be reached. Until more definitive data become available, these authors are inclined to recommend the solid-curve cost data shown on Figure 12.

The accounting and financing methods for LNG ships are varied and complex. It was virtually impossible to obtain a common accounting base for the data presented in Table 18. However, a common base was maintained where sufficient data were available to permit cost adjustments. The fabrication cost of LNG vessels was increased by 10% to account for the higher cost of liquid hydrogen equipment.

The replacement cost of the NASA liquid hydrogen barge[44] was estimated at \$1.58 × 10^6 and an annual charge rate of 26%[48] was assessed. Actual operating costs were used to assess freight charges (barge operator labor plus tug fee) for this barge. Reproduction of the LNG barge[46,47] for liquid hydrogen service was estimated at \$7.39 × 10^6, and an annual charge rate of 26% was assumed. Total operating costs were taken at \$9900/day – \$6400/day for ownership and \$3500/day for freight charges. The cost data of Backhaus and Janssen[43] were simply corrected to account for inflation and the higher cost of hydrogen equipment. The economic data of Donovan[48] and Soedjanto et al.[49] were similarly adjusted. An annual charge rate of 26% was used to adjust the data of Soedjanto et al. All barges were assumed to operate 300 days/year, and all ships were allowed to operate 345 days/year. Capital costs of the large liquid hydrogen ships were estimated at \$86 × 10^6 (11 × 10^6 lb [5 × 10^6 kg] hydrogen capacity[48]), \$128 × 10^6 (19.4 × 10^6 lb [8.8 × 10^6 kg] hydrogen capacity[48]), and \$135.30 × 10^6 (19.4 × 10^6 lb [8.8 × 10^6 kg] hydrogen capacity[49]).

1.4.9. Metal Hydride by Highway and Railway

The use of metal hydrides is in its infancy; design data are not readily available for economic analysis. One hydride storage design[50] has been published and will form the basis of the economic discussion in this section. A safe ore is a prerequisite for transportation of hydrogen by highway or railway. Iron-titanium alloy is believed to fulfill the general safety requirements,[51] has desirable charge-discharge pressure-temperature characteristics,[50] has the potential of being relatively inexpensive[52] and readily available, and is the chosen ore in our reference design.[50] These reference data are taken from a prototype stationary storage design and are not necessarily optimum for portable containers; however, they should provide a reasonably good indication of the potential economics of transporting hydrogen as a metal hydride.

The reference design data for the hydride reservoir are listed below.

- Alloy: Iron-titanium ($FeTiH_{0.2}$ to $FeTiH_{1.5}$)
- Container: 12-in. (0.3048-m) diameter, schedule 20 pipe, 6.5 ft (1.98 m) long, 316 stainless steel
- H_2 storage capacity: 14 lb (6.35 kg)
- H_2 delivery capacity: 12 lb (5.44 kg)
- Gross weight of storage reservoir: 1241 lb (562.9 kg)
- Weight of alloy: 879 lb (398.7 kg)
- Weight of hydrogen header, heat exchanger, and shell (container): 348 lb (157.9 kg)
- Volume available for alloy: 4.66 ft^3 (0.132 m^3)
- Unhydrided alloy volume: 4.28 ft^3 (0.121 m^3)
- Density of alloy bed: 206 lb/ft^3 (3300 kg/m^3)
- Density of hydride: 353 to 398 lb/ft^3 (5655 to 6375 kg/m^3)
- Alloy void fraction: 0.492
- Heat exchanger surface area: 18.8 ft^2 (1.75 m^2)
- H_2 header surface area: 5.87 ft^2 (0.55 m^2)
- Hydriding (charging) rate: 1.5 lb/hr (0.68 kg/hr)
- Dehydriding (discharging) rate: 1.0 lb/hr (0.45 kg/hr)
- Design pressure: 633 psig (4.36 MPa)
- Hydriding pressure: 500 psig (3.45 MPa)
- Dehydriding pressure: ~1 to 10 atm
- Heat required to form hydride: 6700 Btu/lb (15.57 MJ/kg)
- Heat required to dehydride: 6700 Btu/lb (15.57 MJ/kg)
- Heat exchange fluid: water at ~3.3 gal/min (208.2 cm^3/sec) and 66°F (292 K).

All materials used in this hydride storage container were 300-series stainless steels; however, it is assumed herein that lower-cost steels, such as those used in high-pressure hydrogen cylinders, would be used in future commercial applications. Using the foregoing data, we can derive some interesting cost figures. Perspective is obtained by comparing these cost figures with those previously derived for transport of compressed hydrogen gas by highway tube trailer and railroad tank car.

The effective hydrogen storage density of the Fe-Ti ore is $12 \div 4.66 = 2.58$ lb/ft^3 (41.33 kg/m^3), and the storage density of room temperature hydrogen gas at 2400 psia (16.55 MPa) is about 0.77 lb/ft^3 (12.33 kg/m^3). Consequently, a significant reduction in container volume is possible with hydride storage. The ratio of system weight to delivered product weight is $1241 \div 12 \approx 103$ for the hydride system, as compared to 200 for the railroad tank car (compressed gas) and 95 for the highway tube trailer (compressed gas). The additional weight of structural cabinetry and running gear would increase this ratio for the hydride system; however, it appears that the hydride transporter could be fairly weight-competitive with the highway tube trailer and lighter than the railroad tank car. Thus, we would anticipate that freight rates for hydride transportation would be comparable and competitive with those for transportation of compressed hydrogen gas. The hydride freight rates may even be lower due to the reduced volume and larger delivery capacity of the hydride system.

An estimate of capital costs will complete our comparison of transportation schemes for gaseous hydrogen. First, we will estimate the cost of an Fe-Ti highway transporter. Scaling up the Fe-Ti reference design data and using the cost data of Guthrie[53] for high-pressure shell-and-tube heat exchangers, we obtain an estimated cost of $42,500 for a hydride reservoir that will deliver 115,000 SCF (272.6 kg) of hydrogen gas. Adding $3000 for the trailer running gear; $3000 for miscellaneous valves, plumbing, and structural materials; and $22,000 for Fe-Ti ore (taken at $0.50/lb [1.10/kg]), the total cost of the hydride highway transporter is estimated at $70,500. If the ore costs $1.00/lb ($2.20/kg), the initial cost of this vehicle would be estimated at $92,500. These estimated costs are near-competitive with the current replacement costs of highway tube trailers (~$70,000). Repeating this approach for Fe-Ti railway transporters (210,000 SCF [497.7 kg] hydrogen capacity), we estimate capital costs ranging from $132,000 to $172,000, depending upon the cost of ore. These cost estimates are lower than current replacement costs for railroad tank cars (~$200,000).

From these preliminary estimates, it appears that hydride transportation of hydrogen is feasible, and that freight and capital charges are likely to be competitive with those of existing methods for transporting gaseous hydrogen by highway and railway. Hydride transportation offers some potential advantages: 1) smaller container volumes and larger delivery capacities, 2) lower operating pressure and thus lower compressor costs, and 3) possibly lower unit transportation charges.

No additional cost analysis effort is justified until more specific design and economic data become available. Conceivably, other hydrides may ultimately offer better economic advantages. Other alloys were not considered herein because they were believed to possess one or more adverse characteristics related to safety, availability, initial cost, operating pressures, and temperatures, heat of hydride formation, susceptibility to contamination, and lack of reliable design data.

1.4.10. Hydrogen by Airline

Shipment of liquid hydrogen by airplane is not allowed under current DOT regulations.[54,55] Compressed hydrogen gas may be transported by cargo aircraft in DOT-approved[54] cylinders with a maximum charge of 300 lb (136 kg) of hydrogen per cylinder. Some commercial air carriers are currently proposing that the ban on transportation of flammable or toxic materials on passenger aircraft be extended to include cargo aircraft.

1.4.11. Summary of Hydrogen Transmission Costs

As has been demonstrated throughout this section, it is quite apparent that transmission costs are system dependent. Terminal facilities and terminal costs vary appreciably with the fluid state and fuel application, e.g., the need for high-pressure gas, low-pressure gas, or liquid can significantly alter terminal costs. All of these costs must be assessed in any complete fuel system; however, for convenience and simplicity, we have estimated transmission costs excluding terminal charges. As indicated earlier, these charges are sometimes included in the production and distribution cost increments. Table 19 summarizes the transmission costs developed in this section. Selection of an optimum mode of transportation will require careful definition of the fuel system requirements.

1.5. HYDROGEN STORAGE COSTS

In this section, we consider the cost elements

TABLE 19

Summary of Hydrogen Transmission Costs

Mode of transmission	Estimated unit transmission cost	
	$/10⁶ Btu-mi	$/kg-km
Overland gas pipeline	0.00043–0.00071	0.000036–0.000059
Underwater gas pipeline	0.00132–0.01000	0.000110–0.000836
Compressed gas by highway tube trailer	0.041–0.068	0.0034–0.0057
Compressed gas by railroad tank car	0.029–0.037	0.0024–0.0031
Compressed gas by barge or ship	0.035–0.065	0.0029–0.0055
Liquid pipeline	0.0096–0.0280	0.00080–0.00234
Liquid by highway tank trailer	0.0059–0.0075	0.00049–0.00063
Liquid by railroad tank car	0.0026–0.0073	0.00022–0.00061
Liquid by barge or ship	0.00037–0.0057	0.000031–0.00048

Note: All rates based on one-way mileage, excluding terminal costs, right-of-way costs, etc. (see Tables 10 through 18 and Figures 10 through 12 for specific assumptions and technical details).

for various modes of storing hydrogen: 1) compressed gas in underground reservoirs (e.g., depleted oil or gas fields), 2) compressed gas in metal containers, 3) metal hydride, and 4) liquid.

The optimum storage method must be determined on the basis of cost, quantity stored, desired purity level, and the desired temperatures, pressures, and flow rates of the incoming and outgoing hydrogen.

The costs presented in this section, excluding underground storage, include only the storage vessel; the costs for ancillary equipment, such as compressors and transmission piping are given elsewhere in this chapter.

1.5.1. Underground Storage

Underground storage in depleted oil or gas fields and in aquifers is an attractive means of storing large quantities (10^{10} SCF) of natural gas.[58] Underground storage is used for daily, weekly and seasonal load leveling. A recent paper by Walters[59] evaluated the feasibility of underground storage of hydrogen. Although cost factors vary, depending upon the location and nature of each underground reservoir, Walters concludes that total capital costs for storing hydrogen underground will be between \$3 and \$6/10^6 Btu (\$0.40 to \$0.80/kg); total operating costs are estimated at \$1 to \$3/10^6 Btu (\$0.13 to \$0.40/kg) of gas handled. These costs are three to four times greater, on an equivalent energy basis, than underground natural gas storage. Hydrogen storage is more expensive than natural gas storage because the volumetric heating value of hydrogen is one third that of natural gas; therefore, hydrogen will require three times as much storage volume for a given energy-storage capacity. Also, the volumetric delivery rate for hydrogen must be three times that of natural gas for equivalent energy flow.

Another important factor will be the cost of the hydrogen used as a "cushion gas" – the gas which must remain in the reservoir to maintain the desired delivery rate. Current reservoirs have a cushion gas-to-deliverable gas ratio of ~0.5 to 2.0 with an average ratio of 1.41.[58] If hydrogen is expensive, reservoirs requiring high cushion gas-to-working gas ratios will require a high capital cost for the hydrogen cushion gas.

Conversion of existing underground natural gas reservoirs to hydrogen service appears feasible. The major change will be the replacement of natural gas compressors with hydrogen compressors. If hydrogen is stored at high pressure, some of the high-pressure pipelines may have to be replaced with materials resistant to hydrogen embrittlement.

1.5.2. Aboveground Storage

Hydrogen can be stored as a gas aboveground in low-pressure gas holders or in pressure vessels. Low-pressure gas holders are too expensive to be considered for any type of storage other than temporary or large-quantity storage. However, high-pressure cylinder storage offers a relatively simple and inexpensive means of storing small quantities of hydrogen.

Table 20 gives representative costs for storing compressed hydrogen gas in steel cylinders.[9] The cost of compressing the hydrogen to the desired storage pressure is given in Section 1.3.2.

1.5.3. Metal Hydride Storage

To estimate the cost of storing hydrogen as a metal hydride, we use the same approach outlined

TABLE 20

Cost of Storing Hydrogen Gas in Steel Cylinders

	Maximum pressure		Storage capacity		Costs per cylinder, \$		
Cylinder dimensions	psia	MPa	10^6 Btu/cyl	kg/cyl	Cylinder	Mounting	Manifolding
24 in. (0.61 m) O.D.	500	3.45	1.23	9.14	3300	540	250
40 ft (12.2 m) length	1295	8.93	2.80	20.81	3500	540	275
24 in. (0.61 m) O.D.	2495	17.20	2.39	17.76	3900	540	300
20.5 ft (6.25 m) length							

Note: Estimated installation cost is 30% of capital cost; estimated O & M cost is 3% of installed capital cost per year.

From Electric Power Research Institute Report No. 320-1, August 1975.

in Section 1.4.9. The two major cost items are the hydride container, which includes a heat exchanger, and the hydriding metal. The hydride system considered in Section 1.4.9. stores 12 lb (5.44 kg) of hydrogen and delivers hydrogen at 1 lb/h (0.454 kg/h). If we retain the heat exchanger design of this system, we can expect a hydrogen discharge rate that varies linearly with the quantity of hydrogen stored, i.e., (1 lb/h) ÷ (12 lb) = 0.0833 h^{-1}. This design provides an hourly delivery of 8.33% of the maximum quantity of deliverable hydrogen that is stored in a hydride bed of any desired size. Figure 13 gives the installed cost for a container module (with heat exchanger) as a function of storage capacity. The estimated costs are based on a heat exchanger design that provides an area of 1.57 ft^2 per pound of hydrogen delivered (0.32 m^2/kg); these costs were obtained by inflating the cost data of Guthrie.[53] The amount of alloy required is 1200 lb/10^6 Btu of deliverable hydrogen (73.3 kg ore per kg hydrogen). Figure 14 shows how the installed cost of a hydride storage system varies with the quantity of deliverable hydrogen stored; the curve assumes that the iron-titanium costs $1.00/lb ($2.20/kg) and that the hourly discharge rate is 8.33% of the storage capacity. These costs do not include compressors, purifiers, or the auxiliary heat exchange equipment required to cool and heat the hydride bed during charging and discharging operations. The operating and maintenance costs for the hydride system should be low – about 1% of the installed capital cost per year.

One should be aware that these cost data are preliminary cost estimates based on only one hydride system design. As the hydride storage technology advances, more meaningful cost data should become available. Salzano et al.[62] have recently provided cost estimates for a specific hydride storage design of the type considered here – 1700 × 10^6 Btu storage capacity (10 storage units at 170 × 10^6 Btu per storage module) with a discharge rate of 10% of storage capacity per hour.

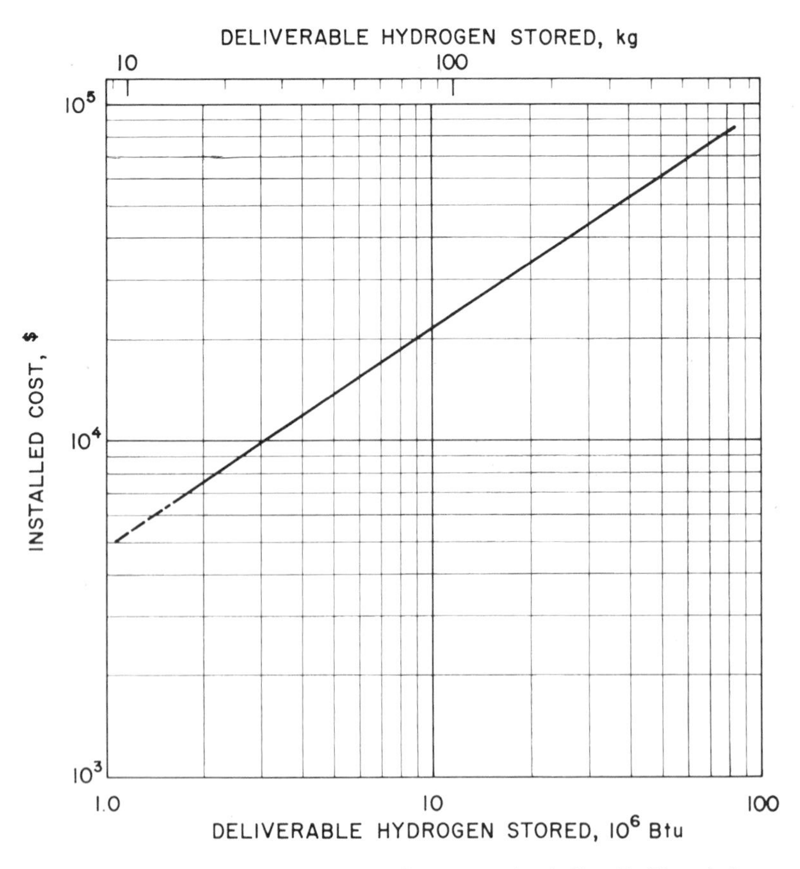

FIGURE 13. Installed cost of hydride storage (excluding Fe-Ti ore) for a discharge rate of 8.33% of storage capacity per hour.

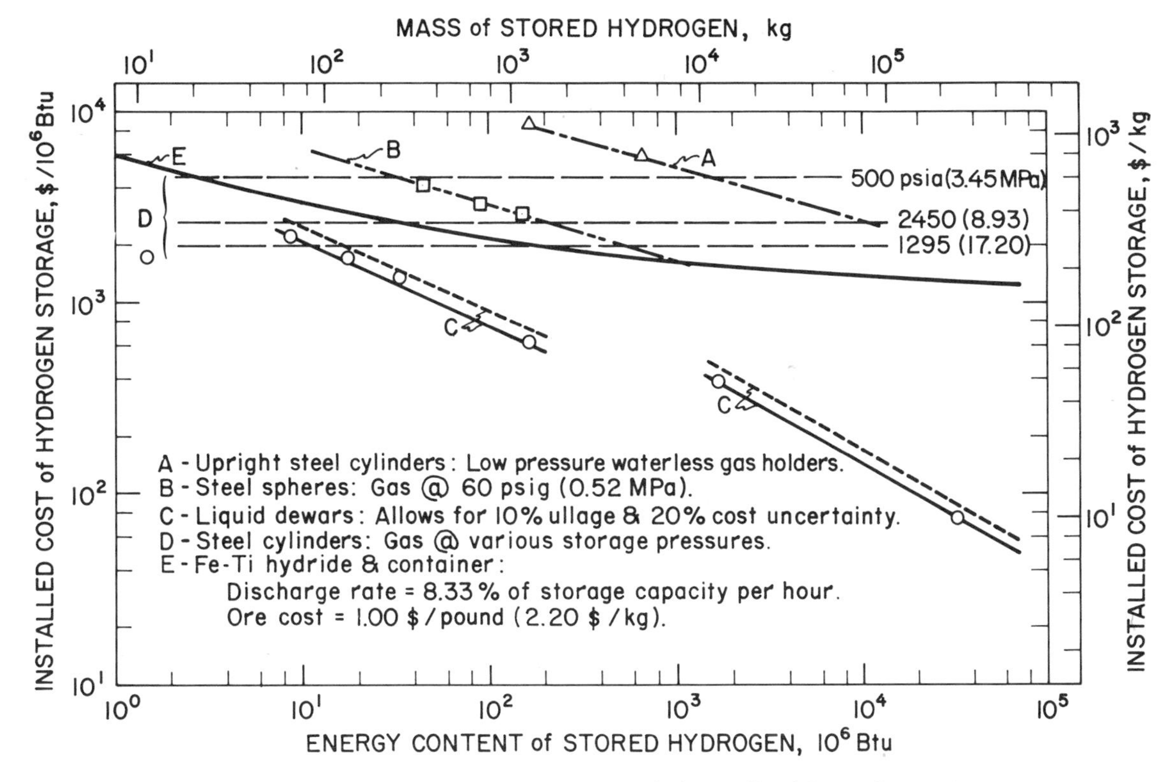

FIGURE 14. Cost of storing hydrogen in the compressed gas, hydride, and liquid forms. Costs and storage quantities are based on deliverable hydrogen stored.

These costs[62] include purification and auxiliary equipment costs and exceed those shown on Figure 14 by about 35%.

1.5.4. Liquid Hydrogen Storage

Next, we consider cost data for liquid hydrogen dewars; the cost of liquefaction is given in Section 1.3.2. Table 21 gives cost data, and Figure 14 shows the estimated installed capital costs for liquid hydrogen dewars. The unit storage costs are based on installed costs and allow for 10% ullage volume.

Costs for the two largest dewars are approaching a minimal value[61] because they are based upon a low labor-cost area (Gulf coast), an aluminum inner tank wall, and a minimal amount of insulation (2½ feet of perlite). In installations where the constraints cited above are not valid, a large liquid hydrogen dewar could cost 20% more than the values given in Table 21.

1.6. EXAMPLES OF HYDROGEN SYSTEM COST ANALYSES

In the costing examples that follow, we deliberately chose hydrogen energy systems that illustrate the economy of scale. The first example, for a solar-hydrogen energy system, is considered one of the smallest feasible systems for future applications. The second example, for supply of hydrogen to an international airport, is one of the largest possible applications. In both examples, we avoid specific accounting procedures and simply use our developed cost data to estimate the cost of delivered hydrogen energy.

1.6.1. A Solar-hydrogen Energy System

Many energy experts have come to believe that geophysical energy sources such as solar, wind, oceanthermal gradients, etc. are potentially the best alternatives to the waning fossil fuels because they are abundant and nonpolluting. For example, a solar collector having an area of about 110 m^2 could supply all of the heating, cooling, hot water, and electrical needs of a single household in Boulder, Colorado; a wind turbine of 30 m^2 wind intercepting area could provide all of the electrical needs of that household. However, the difficult problems caused by the diffuse and intermittent nature of the geophysical energy sources have prevented their widespread acceptance. This example outlines an energy system that combines

TABLE 21

Cost of Installed Liquid Hydrogen Dewars

Capacity				Maximum operating pressure		Unit storage cost[a]	
gal	m³	Capital cost ($)	Boiloff rate (%/day)	psia	MPa	$/10⁶ Btu	$/kg
46.2	0.175	1,720[b]	1.50	25	0.17	1,731	233
264	1.00	12,600[b]	1.25	25	0.17	2,206	297
528	2.00	19,500[b]	1.00	25	0.17	1,707	230
1,000	3.79	29,700[b]	1.00	25	0.17	1,373	185
5,000	18.93	66,200[b]	0.50	25	0.17	612	82.3
50,000	189.3	625,000[c]	0.20	45	0.31	385	51.8
1,000,000	3785.7	2,450,000[c]	0.06	45	0.31	75.5	10.2

Note: O & M cost estimated at 2% of installed capital cost per year.

[a]This cost allows for 10% ullage volume.
[b]This cost excludes delivery and installation;[60] installation costs are estimated at 50% of these f.o.b. capital costs.
[c]Daniels;[61] this is an installed capital cost.

solar collectors and wind turbines to form a total energy system that is independent of outside utilities for a community of 1000 or more residences. This is made possible by combining solar and wind energy collectors to take advantage of the tendency of these two sources to complement each other, and the use of hydrogen as an energy storage medium. This system looks ahead to a time when such a community could be an interacting part of an integrated hydrogen-electric economy. The systems used to collect the energy, and to produce and store the hydrogen, are complex; therefore, the implementation of such a plan would only be economical on a scale of 1000 or more residences, for a compact housing complex of 1000 dwelling units, or for a large shopping center or industrial facility.

The solar-hydrogen system discussed herein has numerous advantages over more conventional solar energy systems: (1) the system has redundancy for supplying energy for space heating, cooling, hot water, and electricity, so it will provide the required reliability for utilities; (2) the system is self-contained and does not depend on energy sources external to the system during normal operation; and (3) if the solar collectors or wind turbines fail, emergency energy can be supplied from an external source in the form of hydrogen.

We divide the remaining discussion of this example problem into two categories: a technical review of the pertinent design features of the reference system, and a cost analysis of the reference system as designed. The technical features, although essential to the preparation of a cost analysis, are briefly reviewed as they have been outlined and discussed in detail elsewhere.[63]

1.6.1.1. Technical Review

Solar heating and cooling of buildings, as well as generation of electricity by photovoltaic devices, solar thermal processes, and wind turbines, have been demonstrated; however, commercialization has been deterred because capital investments are considered too high and because engineers have not yet learned to cope with the capricious nature of these geophysical energy sources. To some degree, these two deterrents are really one, since much of the investment in proposed solar or wind power plants is due to oversizing of systems to meet peak demands, and the need for auxiliary-fueled power and heating plants when the solar/wind sources are inactive.

From the viewpoint of the electric utilities industry, solar/wind partial electric power generation and heating backed up by utilities power is not an attractive prospect.[64] To back up partial solar or wind power systems, these utility companies would still need an installed capacity sufficient for the peak loads; furthermore, the disparity between peak and nonpeak loads would be worsened. The deleterious effect on power plant efficiency and the increased operating expense passed on to the customers would certainly detract from the real value of a solar/

wind power installation. Clearly, the transition from partial heating or cooling to a totally independent geophysical energy supply would be a significant advancement. One step toward this goal is to combine solar collectors and wind turbines into a hybrid system. Since winds prevail in the winter months and during the night when insolation is lacking, the combination tends to eliminate some of the lapses which would occur in either source by itself. An additional prerequisite to a totally independent energy system is a means of storing energy for long periods of time (6 months or longer) and a means of recovering that energy efficiently. A comparison of various methods of storing solar energy is given in Figure 15.

Heat received from solar collectors may be stored directly as sensible or latent heat in various phase-change materials. Water is one of the best media for storage of low-grade heat because of its high sensible heat, but it is limited to a temperature of 100°C unless it is pressurized, and large pressure vessels are expensive. Other sensible heat storage materials such as rocks require large volumes, and suitably processed alumina is expensive. Phase-change materials are generally corrosive and may require special encapsulation, which again makes them expensive. They also present problems of undesirable supercooling before solidification occurs and of stability of composition. Another problem with direct storage of heat for eventual generation of electricity is that only 20 to 40% of the stored heat can be converted to electrical energy; the rest will be wasted or must be re-stored as low-grade heat. The alternative of storing electrical energy in batteries is not appealing because of the expense, poor lasting qualities, losses, and poor energy density of batteries. The superflywheel may eventually perform better than batteries, but there are problems of energy extraction, etc.

Hydrogen as an energy storage medium has many outstanding advantages:

1. Hydrogen may be produced from electrolysis of water with high efficiency of conversion from electrical to chemical energy (80% or better);
2. Hydrogen may be converted back to electrical energy with high efficiency (55% in fuel cells, 40% in thermal cycles);
3. Hydrogen is easily stored and is transportable. Compactness is gained when the hydrogen is stored in the liquid form or in metal hydrides as shown in Figure 15;

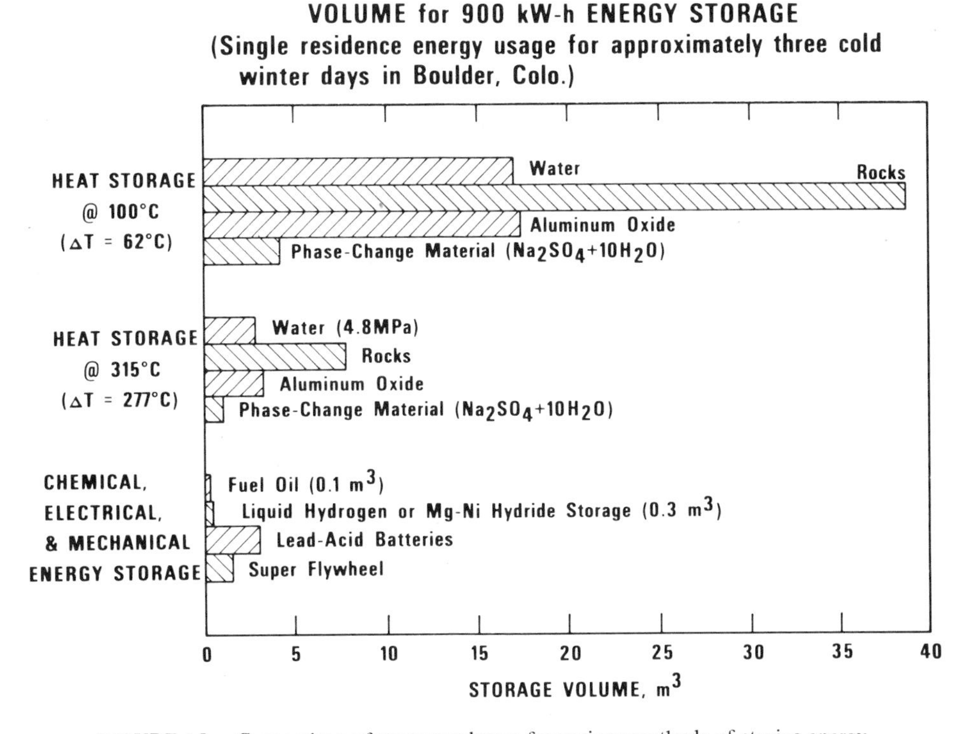

FIGURE 15. Comparison of storage volumes for various methods of storing energy.

4. When burned as a fuel or used in a fuel cell, the products of hydrogen combustion are nonpolluting. This characteristic also allows for almost perfect recovery of the heat of combustion when hydrogen is burned to produce heat. Since the product of combustion is simply water vapor, the combustion product may be condensed, the water recovered, and the higher heating value of the fuel realized.

In the cost analysis that follows, we examine three methods of storing hydrogen — liquid, compressed gas in steel cylinders, and metal hydride.

The solar-hydrogen system is shown schematically in Figure 16; it consists of a concentrating solar collector and a wind turbine, both supplying d.c. electrical power to an electrolysis unit. The term "Solar collector" as defined herein includes a light-concentrating device, and a means of collecting heat and transferring it for direct use or storage. The solar collector supplies superheated vapor of a suitable heat transfer fluid to a Rankine cycle turbine. The thermal cycle for the turbine operates in two modes. In the winter months, when space heating is required, heat from the turbine exhaust condenser is recovered and stored in water at 100°C for space heating. The resulting rise in turbine exhaust temperature decreases thermal cycle efficiency, but that decrease is more than compensated by the recovery of a large amount of low-grade heat. In the summer, when space heating is not needed, the turbine exhaust condenser is operated at the lowest possible temperature, thereby increasing the thermal cycle efficiency and turbine power. Additional turbine power is required in summer for direct drive of vapor compressors for air conditioning. The storage of heat from the solar collector at high temperature in aluminum oxide enables the turbine to operate from stored heat as required to eliminate short term lapses in power; however, preliminary calculations indicate it is impractical to store sufficient high-temperature heat to bridge periods of a week without sunlight or wind.

Electrical power generated by the solar thermal cycle and wind turbine is used directly as needed and the surplus power is used to electrolyze water.

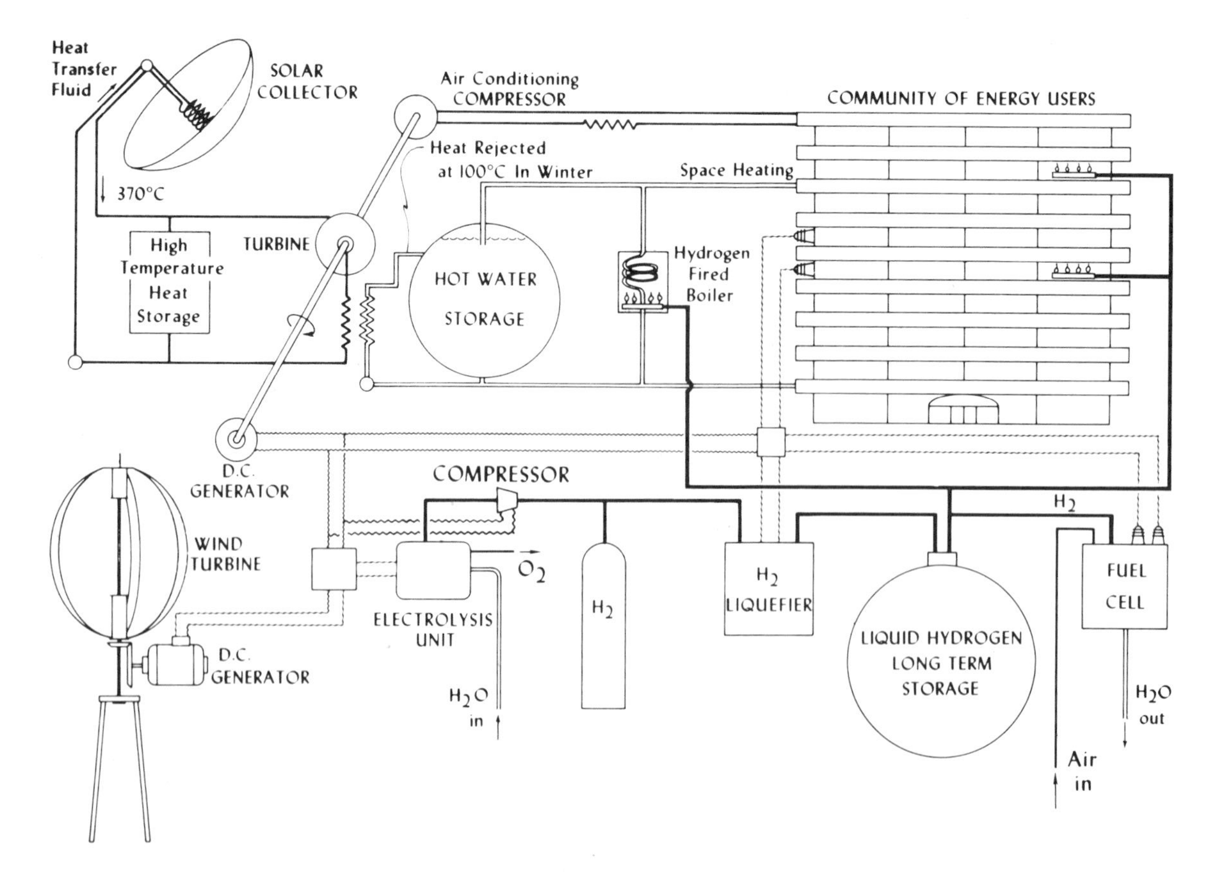

FIGURE 16. A solar-hydrogen-powered community.

The hydrogen gas compressor operates only when sufficient electricity is being generated to supply the electrolyzer. Short-term storage is provided for the intermittently produced hydrogen in pressurized cylinders, and long-term storage is provided by one of the storage options previously outlined. The hydrogen liquefier and liquid storage dewar, as shown in Figure 16, are replaced by high-pressure cylinders or metal hydride containers to separately evaluate the economics of each of these hydrogen storage options. The stored hydrogen supplies electrical power by fuel cell conversion; it may also be burned in oven, range, and dryer appliances and in a hot-water boiler to supplement the space heating that is ordinarily supplied by rejected turbine heat.

While the hydrogen energy storage methods employed in this example are independent of the energy supply in the adjoining community, the system would function best as an interacting part of a general hydrogen economy. A widespread network of cooperating hydrogen suppliers and consumers would provide redundancy in cases of equipment failure or localized deficiencies in production. The size of such a system is not limited by the characteristics of the hydrogen storage concept; however, there are considerations which would seem to narrow the size of the optimum *total energy* system for a singular installation to something smaller than a large city and something larger than an individual dwelling. The first of these considerations is the desirability of using heat rejected from the thermal electric power cycle for space heating in winter and tap-water heating. This is an important consideration, since this rejected heat may amount to 70 to 80% of the total output of the solar collector. In many regions of the U.S., this ratio of thermal to electrical power (75:25) roughly matches the energy demand in winter. The use of the rejected heat would not be possible if the distance from the power plant to the user were too great; this would tend to limit the size of the area served by a single installation to a small community or a small section of a city. For very large, remote power plants, the emphasis would probably be placed on obtaining the best possible thermal cycle efficiency by operating at the lowest possible turbine exhaust temperature. The sacrifice of the rejected heat would have to be weighed against the other advantages of very large-scale installations. A solar-hydrogen system for very small installations, such as single-family dwellings, is not practical because of the technical complexities and attendant high costs of small units.

1.6.1.2. Cost Analysis

Commercial fuel cells and concentrating solar collectors are not yet available, and the costs assumed herein are somewhat speculative; however, most other equipment component costs are reasonably well known, and some are documented in previous sections of this chapter. The estimated capital costs and capacities of the various sytem components are given in Tables 22 through 24 for the three hydrogen storage options considered. Cost estimates for other solar-hydrogen energy concepts are given in Eisenstadt and Cox[65] and McCulloch et al.[66]

The reader can readily justify the component costs of the hydrogen equipment by referring to the cost data previously developed in this chapter. By comparing the capital cost data in Tables 22 through 24, it is clear that the liquid hydrogen storage system is the most economical storage option for relatively large-scale long-term storage of hydrogen. Although the liquid storage system requires the largest solar collector area, the cost of compressed gas or hydride storage is prohibitive for this application.

To compute the total cost of delivered energy, we will assume a FCR of 0.13 and commit 2% of the capital investment for IRRI. An O & M charge of 3% of capital cost is assessed on all system components except the compressed gas cylinders and hydride storage equipment. The latter two components are charged 1% for O & M. The resultant annual charge rates are 18% of the total capital investment for the solar-hydrogen plant using liquid hydrogen storage and slightly lower for the two systems using gaseous hydrogen storage. The installed costs of solar collectors vary, and the resulting annual operating costs for each solar-hydrogen system are computed and divided by the annual energy output of these systems (66.9×10^6 kWh per year). The delivered energy costs are plotted in Figure 17 as a function of solar collector capital costs; these costs include property taxes, income taxes, and insurance. Also, for purposes of comparison, the current operating costs are shown for similar households using electricity only or electricity plus natural gas.

Figure 17 reveals that the delivered energy costs are quite sensitive to the cost of the solar collector, and that the energy costs for the

TABLE 22

Size and Cost of Solar-hydrogen Energy System Components – Based on Liquid Hydrogen Storage and System Components Sized to Service 1,000 Dwelling Units

Equipment component	Size	Capital Cost
Wind turbine generator	2,500 kW	$\$2.75 \times 10^6$
Heat engine generator	15,500 kW	2.17
Insulated high-temperature heat storage tank (@ 370°C)	4,000 m^3	0.42
Insulated hot-water storage tank (@ ~100°C)	11,500 m^3	1.98
Al_2O_3	15.5×10^6 kg	1.15
Electrolyzer (@ 3.04 MPa)	227 kg/h	3.49
H_2 gas compressor (3.04 MPa to 8.96 MPa)	227 kg/h	0.08
H_2-air fuel cell	2,500 kW	0.73
Liquefier	243 kg/h	1.89
Liquid H_2 dewar	3,000 m^3 (190,000 kg H_2)	1.94
H_2 gas storage (@ 8.96 MPa)	900 kg H_2	0.19
Air conditioner	2,000 ton	0.14
Wiring and auxiliary electrical power equipment	–	0.46
Yard and distribution piping and auxiliary heat exchangers	–	0.78
Engineering, field connections, and checkout	–	2.54
Alarms, controls, and safeguards	–	0.10
Capital investment excluding solar collector		$\$20.81 \times 10^6$
Solar collector[a]	110,000 m^2	–

[a]Solar collector costs are taken as variable; see Figure 17.

TABLE 23

Size and Cost of Solar-hydrogen Energy System Components – Based on Compressed Hydrogen Gas Storage and System Components Sized to Service 1,000 Dwelling Units.

Equipment component	Size	Capital cost
Wind turbine generator	2,100 kW	$\$2.31 \times 10^6$
Heat engine generator	12,837 kW	1.80
Insulated high-temperature heat storage tank (@ 370°C)	4,000 m^3	0.42
Insulated hot-water storage tank (@ ~100°C)	11,500 m^3	1.98
Al_2O_3	15.5×10^6 kg	1.15
Electrolyzer (@ 3.04 MPa)	227 kg/h	3.49
H_2 gas compressor (3.04 MPa to 8.96 MPa)	227 kg/h	0.08
H_2-air fuel cell	2,500 kW	0.73
H_2 gas storage (@ 8.96 MPa)	190,900 kg H_2	40.15
Air conditioner	2,000 ton	0.14
Wiring and auxiliary electrical power equipment	–	0.42
Yard and distribution piping and auxiliary heat exchangers	–	0.65
Engineering, field connections, and checkout	–	2.10
Alarms, controls, and safeguards	–	0.10
Capital investment excluding solar collector		$\$55.52 \times 10^6$
Solar collector[a]	106,427 m^2	–

[a]Solar collector costs are taken as variable, see Figure 17.

TABLE 24

Size and Cost of Solar-hydrogen Energy System Components – Based on Fe-Ti Hydride Storage and System Components Sized to Service 1,000 Dwelling Units

Equipment component	Size	Capital cost
Wind turbine generator	2,270 kW	$2.50 × 10^6$
Heat engine generator	13,000 kW	1.82
Insulated high-temperature heat storage tank (@ 370°C)	4,000 m^3	0.42
Insulated hot-water storage tank (@ ~100°C)	11,500 m^3	1.98
Al_2O_3	15.5 × 10^6 kg	1.15
Electrolyzer (@ 3.04 MPa)	227 kg/h	3.49
H_2-air fuel cell	2,500 kW	0.73
H_2 purifier	227 kg/h	0.11
Fe-Ti hydride storage	206,350 kg H_2	36.03
Air conditioner	2,000 ton	0.14
Wiring and auxiliary electrical power equipment	–	0.42
Yard and distribution piping and auxiliary heat exchangers	–	0.65
Engineering, field connections, and checkout	–	2.10
Alarms, controls, and safeguards	–	0.10
Capital investment excluding solar collector		$51.64 × 10^6
Solar collector[a]	107,856 m^2	–

[a]Solar collector costs are taken as variable; see Figure 17.

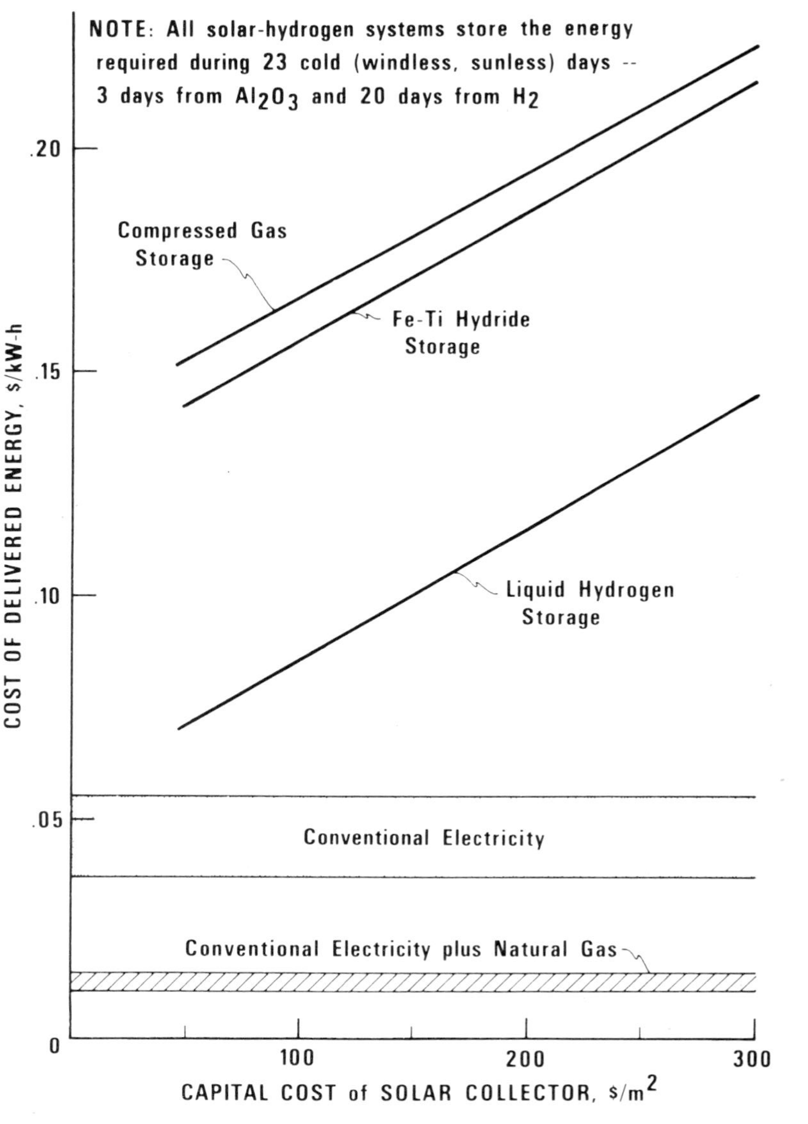

FIGURE 17. Cost of energy vs. solar collector capital costs and current costs of conventional energy in Boulder, Colo.

proposed solar-hydrogen systems are all much higher than conventional energy charges. Even if the solar collector were cost-free, these systems would not currently be cost-competitive with conventional energy sources in Boulder, Colorado; however shortages of natural gas and electricity coupled with their escalating costs could alter this argument in the near future. Also, some additional economy of scale could be expected with the liquid hydrogen storage concept, as nearly all of the system components have lower unit costs in larger sizes. Conversely, both of the gas storage concepts can be excluded from further consideration, because the unit storage costs are independent of size (gas cylinder costs are constant, and the hydride storage costs are approaching the unit cost of the Fe-Ti ore).

Using the procedure outlined in this example, one could now embark on an iterative design costing analysis and develop an optimum solar-hydrogen energy system design. The reader is also invited to show that a more conventional solar energy system, using hot rock and battery energy storage, will not lower the cost of delivered energy. Of course, if the customer is willing to settle for less energy storage capacity, the cost of solar or solar-hydrogen energy can be reduced.

1.6.2. An Airport H_2-fuel Supply System

In this example, we will fix the system design and simply compute the cost of delivered hydrogen. Johnson,[45] Sindt,[67] and Ihara[68] have conclusively shown that system economics favor transmission of hydrogen fuel by gas pipeline. The transmission costs presented in Table 19 also support this conclusion. Consequently, we will analyze a system designed to produce hydrogen gas by electrolysis or coal gasification and transmit the gas through a pipeline to the airport where it is liquefied and stored. A schematic of such a system is shown in Figure 18. The electrolyzer is shown situated near the power plant; however, it could be located elsewhere, as we are charging conventional rates ($0.025 /kWh) for electrical energy consumed by the electrolyzer, compressors, and liquefier. It is assumed that electrical power is available or will be made available (where needed) at this rate. The entire nuclear power plant and electrolyzer are replaced by a coal gasification plant dedicated to the production of hydrogen gas in our comparative cost analysis.

Tables 25 and 26 summarize the component size and capital cost data for the coal-hydrogen and electrolytic hydrogen fuel supply systems. The annual operating charges and unit cost of liquid hydrogen delivered to the airport are given at the bottom of Tables 25 and 26. These costs include property taxes, income taxes, and insurance. Again, the reader can easily verify the hydrogen system component costs from the cost data previously presented in this chapter. It is clear that hydrogen gas produced from coal involves less capital investment and lower operating costs than electrolytic hydrogen. Even if electricity were available at a bus-bar cost of $0.005 /kWh, the electrolytic hydrogen would still be more costly than the coal-hydrogen. Water consumption for the coal-hydrogen is about six times that of the electrolytic hydrogen. The coal process requires a great deal of water for purification, although the hydrogen-producing chemical reactions consume less water than is required for electrolysis (on an equivalent hydrogen output basis).

APPENDIX: CONVERSION FACTORS

1. $3414.43 \frac{\text{Btu}}{\text{kWh}}$
2. $\text{HHV} = 61{,}030 \frac{\text{Btu}}{\text{lb}} = 134{,}545.86 \frac{\text{Btu}}{\text{kg}}$
3. $\text{psi} \times 6894.757 = \frac{\text{N}}{\text{m}^2} = \text{Pa}$
4. $\frac{\$}{\text{kg-km}} \times 0.730 = \frac{\$}{\text{lb-mi(U.S. statute)}}$
5. $\frac{\$}{10^6\ \text{Btu-mi}} \times (83603 \times 10^{-6}) = \frac{\$}{\text{kg-km}}$
6. U.S. statute mi $\times$ 1.609344 = km
7. SCF* $\times$ 0.00523 = lb of hydrogen
8. SCF $\times$ 0.00237 = kg of hydrogen
9. SCF $\times$ 0.0003191869 = 10^6 Btu
10. ft $\times$ 0.3048 = m
11. gal $\times$ 0.003785412 = m^3
12. lb $\times$ 0.4535924 = kg

*SCF = a cubic foot of gas @ 1 atm and 60°F (15.5°C).

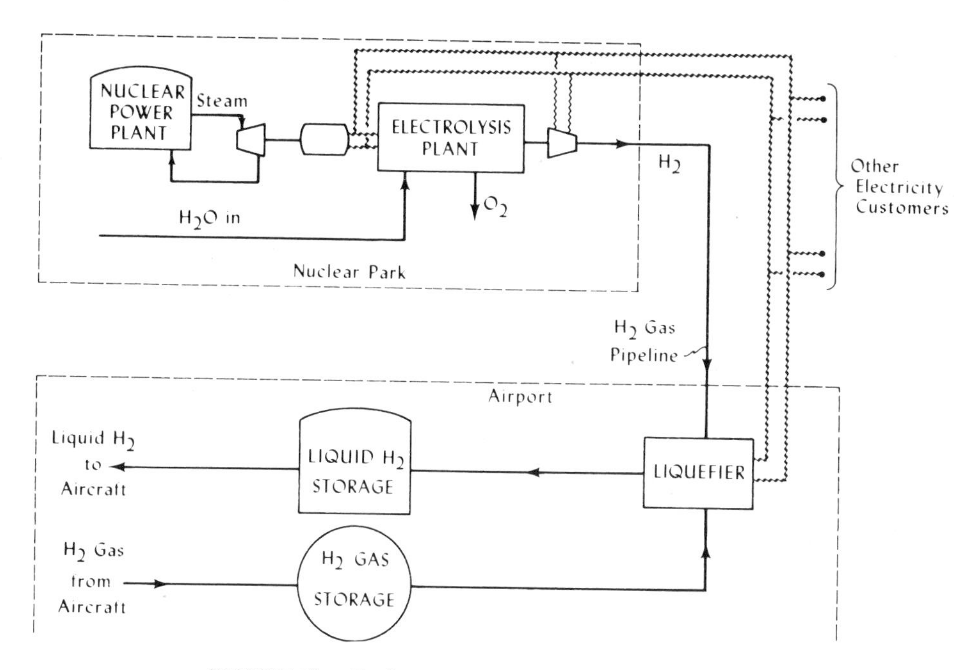

FIGURE 18. H_2-fuel supply system for air transportation.

TABLE 25

Size and Cost of an Electrolytic Hydrogen Fuel Supply System for an Airport – 2.27×10^6 kg/day

Equipment component	Size	Capital cost
Electrolyzer (@ 3.04 MPa)	2.27×10^6 kg/day	$\$1453.9 \times 10^6$
Liquefier (11 units @ 0.227×10^6 kg/day)	2.50×10^6 kg/day	660.0
Liquid H_2 storage (28 dewars @ 10^6 gal each)	106,000 m^3 (6.75×10^6 kg)	68.6
Gas holders (3 holders @ 16×10^6 SCF each)	113,870 kg	48.0
H_2 compressor @ pipeline inlet (3.04 MPa to 5.17 MPa)	2.27×10^6 kg/day	5.5
Yard and distribution piping and auxiliaries	–	77.7
Engineering, field connections, and checkout	–	85.4
Alarms, controls, and safeguards	–	1.1
Capital investment excluding transmission		$\$2400.2 \times 10^6$
Annual charges on I.: @ $0.18 \times \$2400.2 \times 10^6$		432.0×10^6 \$/year
Electricity: 53.8×10^9 kWh/year @ \$0.025 /kWh		1345.0
Raw water: 7.5×10^6 m^3/year @ \$0.0264 /m^3		0.2
Gas transmission: 8.29×10^8 kg/year @ 40×10^{-6} \$/kg-km (100 km overland)		3.3
Total annual charges		1780.5×10^6 \$/year
Unit cost of delivered hydrogen ([1780.5×10^6 \$/yr] ÷ [8.29×10^8 kg/year])		\$2.15 /kg (\$15.98 /10^6 Btu)

13. $\frac{lb}{ft^3} \times 16.0184644 = \frac{kg}{m^3}$

14. $\frac{Btu}{lb} \times 2324.44371 = \frac{J}{kg}$

15. $\frac{gal}{min} \times 0.0000630902 = \frac{m^3}{sec}$

16. $\frac{\$}{10^6\ Btu} \times 0.003414426 = \frac{\$}{kWh}$

17. $\frac{\$}{10^6\ Btu} \times 0.13454586 = \frac{\$}{kg}$

18. J = Btu

TABLE 26

Size and Cost of a Coal-hydrogen Fuel Supply System for an Airport – 2.27 × 10^6 kg/day

Equipment component	Size	Capital cost
Coal gas plant	2.27 × 10^6 kg/day	\$244.8 × 10^6
Liquefier (11 units @ 0.227 × 10^6 kg/day)	2.50 × 10^6 kg/day	660.0
Liquid H_2 storage (28 dewars @ 10^6 gal each)	106,000 m^3 (6.75 × 10^6 kg)	68.6
Gas holders (3 holders @ 16 × 10^6 SCF each)	113,870 kg	48.0
H_2 compressor @ pipeline inlet (3.04–5.17 MPa)	2.27 × 10^6 kg/day	5.5
Yard and distribution piping and auxiliaries	–	77.7
Engineering, field connections, and checkout	–	85.4
Alarms, controls, and safeguards	–	1.1
Capital investment excluding transmission		\$1191.1 × 10^6
Annual charges on I.: @ 0.18 × \$1191.1 × 10^6		214.4 × 10^6 \$/year
Coal: 7.45 × 10^6 tons/year @ \$40 /ton		298.0
Electricity: 10.5 × 10^9 kWh/year @ \$0.025 /kWh		262.5
Raw water: 47.5 × 10^6 m^3/year @ \$0.0264 /m^3		1.3
Gas transmission: 8.29 × 10^8 kg/year @ 40 × 10^{-6} \$/kg-km (100 km overland)		3.3
Total annual charges		779.5 × 10^6 \$/year
Unit cost of delivered hydrogen ([779.5 × 10^6 \$/year) ÷ (8.29 × 10^8 kg/year])		\$0.94 /kg (\$6.99/10^6 Btu)

REFERENCES

1. *Chem. Eng.*, 83(2), 7, 1976.
2. **Stermole, F. J.,** *Economic Evaluation and Investment Decision Methods,* Investment Evaluations Corporation, Golden, Colo., 1974.
3. **Happel, J.,** *Chemical Process Economics,* John Wiley & Sons, New York, 1958.
4. **Siegel, H. M., Kalina, T., and Marshall, H. A.,** Description of gas cost calculation methods being used by the Synthetic Gas-Coal Task Force of the FPC National Gas Survey, paper presented to the Federal Power Commission, Washington, D.C., June 12, 1972.
5. **Katell, S.,** Bureau of Mines, U.S. Dept. of the Interior, Morgantown, W.Va., private communication, 1973.
6. **Tsaros, C. L., Arora, J. L., and Burnham, K. B.,** The manufacture of hydrogen from coal, Paper 751095, presented at the National Aerospace Engineering and Manufacturing Meeting, Los Angeles, Cal., November 1975.

7. **Nahas, N. C.,** Hydrogen manufacturing process, U.S. Patent No. 3,880,987, 1975.
8. **Nahas, N. C.,** private communication, 1975.
9. Utilization of off-peak power to produce industrial hydrogen, Electrical Power Research Institute Report EPRI 320-1, Palo Alto, Cal., August 1975.
10. **Kincaide, W. C.,** Teledyne Energy Systems, Timonium, Md., private communication, 1976.
11. **Fickett, A. P.,** Electrical Power Research Institute, Palo Alto, Cal., private communication, 1975.
12. **Nutall, L. J., Fickett, A. P., and Titterington, W. R.,** Hydrogen generation by solid polymer electrolyte water electrolysis, in *Hydrogen Energy, Part A,* Veziroglu, T. N., Ed., Plenum Press, New York, 1975, 441.
13. **Farbman, G. H.,** Studies of the use of heat from high temperature nuclear sources for hydrogen production processes, Report No. NASA-CR-134918, January 1976.
14. **Gimbrone, G.,** Worthington Compressor Company, Buffalo, N.Y., private communication, 1976.
15. **Strobridge, T. R.,** Cryogenic refrigerators – an updated survey, Natl. Bur. Stand. U.S. Tech. Note 655, 1974.
16. **Voth, R. O. and Daney, D. E.,** H_2 liquefaction: effects of component efficiencies, *Proc. 10th Intersociety Energy Conversion Engineering Conf.,* Institute of Electrical and Electronic Engineers, Inc., New York, 1975, 1356.
17. Survey study of the efficiency and economics of hydrogen liquefaction, Report No. NASA-CR-132631, April 1975.
18. **Parrish, W. R. and Voth, R. O.,** Cost and availability of hydrogen, in *Selected Topics of Hydrogen Fuel,* NBS SP-419, Hord, J., Ed., U.S. Government Printing Office, Washington, D.C., 1975, chap. 1.
19. **Woolley, H. W., Scott, R. B., and Brickwedde, F. G.,** Compilation of thermal properties of hydrogen in its various isotopic and ortho-para modifications, *J. Res. Natl. Bur. Stand.,* 41, 379, 1948.
20. Bureau of Mines, U.S. Dept. of the Interior, Commodity Data Summaries, 1976.
21. **Beghi, G., Dejace, J., Massaro, C., and Ciborra, B.,** Economics of pipeline transport for hydrogen and oxygen, in *Hydrogen Energy, Part A,* Veziroglu, T. N., Ed., Plenum Press, New York, 1975, 545.
22. **Gough, W. C. and Eastland, R. J.,** The prospects of fusion power, *Sci. Am.,* 224(2), 50, 1971.
23. **Flynn, T. M.,** Pilot plant data for hydrogen isotope distillation, *Chem. Eng. Prog.,* 56(3), 37, 1960.
24. **Stuart, A. K.,** Modern electrolyzer technology, paper presented at the 163rd Meeting of the American Chemical Society, Boston, April 9, 1972.
25. **DeBeni, G. and Marchetti, C.,** A chemical process to decompose water using nuclear heat, paper presented at the 163rd Meeting of the American Chemical Society, Boston, April 9, 1972.
26. **Parrish, W. R.,** Recovery of hydrogen liquefaction energy, *Proc. 10th Intersociety Energy Conversion Engineering Conf.,* Institute of Electrical and Electronic Engineers, Inc., New York, 1975, 1352.
27. **Konopka, A. and Wurm, J.,** Transmission of gaseous hydrogen, *Proc. 9th Intersociety Energy Conversion Engineering Conf.,* 405, 1974 (available from the ASME, 345 East 47th St., New York, N.Y. 10017).
28. **Reynolds, R. A. and Slager, W. L.,** Pipeline transportation of hydrogen, in *Hydrogen Energy, Part A,* Veziroglu, T. N., Ed., Plenum Press, New York, 1975, 533.
29. Federal Power Commission, The 1970 National Power Survey, Part I, I-13-7, 1971 (available from the Superintendent of Documents, U.S. Government Printing Office, Washington, D.C. 20402, price $4.00).
30. **Jacobsen, C. R.,** Federal Power Commission, Washington, D.C., private communication, 1975.
31. **Lambert, D. E.,** New developments for offshore construction, *Pipe Line Ind.,* 43(1), 17, 1975.
32. **Gordon, H. W.,** Brown and Root's approach to deep water construction, *Pipe Line Ind.,* 43(1), 32, 1975.
33. Marathon uses reel barge to lay offshore crude line, *Pipe Line Ind.,* 43(3), 45, 1975.
34. **Brown, R. J.,** How to protect offshore pipelines, *Pipe Line Ind.,* 42(3), 43, 1975.
35. **Heerde, W.,** 10-inch bundle pulled to offshore oil terminal, *Pipe Line Ind.,* 42(5), 35, 1975.
36. **Ewing, R. C.,** Pipeline economics, *Oil Gas J.,* 73(33), 75, 1975.
37. **Deason, D.,** Construction projects, *Pipe Line Ind.,* 43(3), 51, 1975.
38. Code of Federal Regulations, Transportation, Title 49, Parts 170–179, U.S. Government Printing Office, Washington, D.C., 1974.
39. **Gerstner, H.,** Bureau of Mines, Amarillo, Texas, private communication, 1975.
40. Code of Federal Regulations, Shipping, Title 46, Part 146.24, U.S. Government Printing Office, Washington, D.C., 1974.
41. The power transmission project – Progress in 1973, Part III – Systems engineering studies, Tech. Note PTP 27, Brookhaven National Laboratory, Associated Universities, Upton, N.Y., 1974.
42. **Voth, R. O. and Hord, J.,** Economics of cryocables, *Int. J. Hydrogen Energy,* 1, 271, 1976.
43. **Backhaus, H. W. and Janssen, R.,** LNG inland transportation with railway tank cars and river-going tankers, *Proc. 4th International Conf. on Liquefied Natural Gas,* Session VI, Paper 2, June 1974 (available from Institute of Gas Technology, 3424 South State St., Chicago, Il. 60616).
44. **Walker, R. D.,** Marshall Space Flight Center (NASA), Huntsville, Ala., private communication, 1975.
45. **Johnson, J. E.,** The economics of liquid hydrogen supply for air transportation, in *Advances in Cryogenic Engineering,* 19, Timmerhaus, K. D., Ed., Plenum Press, New York, 1974, 12.
46. **Randall, G. A., Jr.,** Distrigas Corporation, Boston, private communication, 1975.
47. **Frangesh, N. E. and Randall, G. A., Jr.,** Distrigas LNG barge operating experience, in *Advances in Cryogenic Engineering,* 21, Timmerhaus, K. D. and Weitzel, D. H., Eds., Plenum Press, New York, 1975, 337.
48. **Donovan, L. J.,** Analysis of LNG marine transportation, Report No. MA-RD-900-74040, Vol. II, Appendices, November 1973 (available from National Technical Information Service, Springfield, Va. 22161).

49. **Soedjanto, P., Schaffert, F. W., and Mason, N. C. M.,** Transporting gas-LNG vs methanol, *Proc. 4th Int. Conf. on Liquefied Natural Gas,* Session V, Paper 2, June 1974 (available from Institute of Gas Technology, 3424 South State St., Chicago, Il. 60616).
50. **Strickland, G. and Reilly, J. J.,** Operating manual for the PS & G hydrogen reservoir containing iron titanium hydride, Report No. BNL 50421, February 1974 (available from National Technical Information Service, Springfield, Va. 22161).
51. **Lundin, C. E. and Lynch, F. E.,** The safety characteristics of FeTi hydride, *Proc. 10th Intersociety Energy Conversion Engineering Conf.,* Institute of Electrical and Electronic Engineers, Inc., New York, 1975, 1386.
52. **Salzano, F. J., Ed.,** Hydrogen storage and production in utility systems, Informal Report No. BNL-20040, Brookhaven National Laboratory, Associated Universities, Upton, N.Y., 1975.
53. **Guthrie, K. M.,** Data and techniques for preliminary capital cost estimating, *Chem. Eng.,* 114, 1969.
54. Code of Federal Regulations, Aeronautics and Space, Title 14, Parts 103.7 and 103.9, U.S. Government Printing Office, Washington, D.C., 1974.
55. Official Air Transport Restricted Articles Tariff No. 6-D, Civil Aeronautics Boards No. 82, Sections I, II, and III, Part A, Airline Tariff Publishing Co., Dulles International Airport, Washington, D.C.
56. **Gregory, D. P.,** A Hydrogen-Energy System, Catalog No. L21173, American Gas Association, Arlington, Va., 1973, 7.
57. **Leeth, G. G.,** Energy transmission systems, *Int. J. Hydrogen Energy,* 1(1), 49, 1976.
58. **Grow, G. C.,** The storage of natural gas, in *Am. Chem. Soc. Div. Fuel Chem. Prepr.,* 19(4), 99, 1974.
59. **Walters, A. B.,** Technical and environmental aspects of underground hydrogen storage, *Proc. 1st World Hydrogen Energy Conf.,* Veziroglu, T. N., Ed., University of Miami, Coral Gables, 1976.
60. **Mortenson, T.,** Cryenco, Denver, Colo., private communication, 1975.
61. **Daniels, J. G.,** Chicago Bridge and Iron Company, Denver, Colo., private communication, 1975.
62. **Salzano, F. J., Braun, C., Beaufrere, A., Srinivasan, S., Strickland, G., and Reilly, J. J.** Hydrogen for energy storage: a progress report of technical developments and possible applications, presented at the Energy Storage Conf., Asilomar Conf. Grounds, Pacific Grove, Cal., Feb. 8 to 13, 1976 (also available as BNL 20931, Department of Applied Science, Brookhaven National Laboratory, Upton, N.Y., 1976).
63. **Sindt, C. F.,** Solar energy – liquid hydrogen, in *Selected Topics on Hydrogen Fuel,* NBS SP-419, Hord, J., Ed., U.S. Government Printing Office, Washington, D.C., 1975, chap. 7.
64. **Spencer, D. F.,** Solar energy: a view from an electric utility standpoint, presented at the American Power Conf., Chicago, Il., April 23, 1975.
65. **Eisenstadt, M. M. and Cox, K. E.,** Hydrogen production from solar energy, *Sol. Energy,* 17(1), 59, 1975.
66. **McCulloch, W. H., Pope, R. B., and Lee, D. O.,** Economic comparison of two solar/hydrogen concepts, Report No. SLA-73-0900, Sandia Laboratories, Albuquerque, N.M., 1973.
67. **Sindt, C. F.,** Transmission of hydrogen, in *Selected Topics on Hydrogen Fuel,* NBS SP-419, Hord, J., Ed., U.S. Government Printing Office, Washington, D.C., 1975, chap. 6.
68. **Ihara, S.,** Cost comparison of superconducting cables and hydrogen pipelines for energy transportation, *Proc. 5th Int. Cryogenics Engineering Conf., Kyoto, 1974,* Mendelssohn, K., Ed., IPC Science and Technology Press, London, 1974, 187.

Chapter 2

Hydrogen and the Environment

Chapter 2
HYDROGEN AND THE ENVIRONMENT

John R. Bartlit

TABLE OF CONTENTS

2.1. INTRODUCTION

Many if not most of today's environmental problems are directly associated with the production, transport, storage, and use of energy. Alaskan oil, offshore drilling, the SST, nuclear power, strip mine legislation, coal plant siting, auto emission standards, the Storm King reservoir – each of these energy-related problems has been an environmental "Fifty-Four Forty or Fight" issue for important regional and national interests. Any large-scale substitution of hydrogen for existing energy forms must inevitably have an impact on such issues.

2.1.1. Regulatory Framework

Few citizens, be they engineers, environmentalists, legislators, or laymen, have a practical understanding of the regulation-setting process. Typically 10% of an audience at a public environmental discussion has ever attended a regulatory hearing. Yet the mechanism of regulation setting has a profound influence on regulations themselves, on the environment, and on technologies.

2.1.1.1. Emission Standards vs. Ambient Standards

One of the concepts basic to most clean air and water restrictions is that of emission standards or requirements and ambient air (or water) quality requirements and of the interrelations between emission and ambient requirements. Emission standards are prescribed rules typically stating that an industrial plant of a certain kind and size is permitted to emit a specified weight of a particular pollutant per hour. A coal-fired power plant in New Mexico is permitted to emit 0.05 lbs of particulate matter per hour per million Btus heat input, of which no more than 0.02 lb may be of aerodynamic diameter less than 2 μm in size.

Ambient standards are quantitative specifications of the maximum concentration of each pollutant from all sources permitted in the ambient air or watercourse. Ambient standards generally are not legally enforceable against a pollution source, but are ultimate limits which indicate failure of the overall control program. On an annual average, 60 $\mu g/m^3$ of particulate matter are allowed in the ambient air.

2.1.2. Regulatory Proceedings

Environment is science and technology, but it is more than these. It is also politics and law; these elements also must be understood. People generally imagine that environmental regulations are established through scientific deliberations by highly skilled scientists whose goal is to enlist the best knowledge and advanced techniques available for understanding and elimination, or at least minimizing, environmental damage. In actuality, a regulatory proceeding is a legal proceeding, not a scientific forum. Legal proceedings are adversary proceedings in which scientists are hired as expert witnesses by an interested party for the purpose of defending narrow, specified points fundamental to that party's legal case.

Typically, the adversary parties consist of a government agency or a citizens' group seeking environmental control vs. an environmental polluter seeking to avoid control. Advocates of newer control technologies, of which hydrogen must be considered an example, must present in person the technical and economic facts and arguments favoring the new controls. Competing interests will argue the reverse. The two presentations combined will form a legal record from which a board can make a judgment. By law, the board can consider no alternatives that are not presented and advocated by some party.

2.1.3. Hydrogen and Environmental Standards

The statement most frequently made about hydrogen and the environment is "Hydrogen is a pollution-free fuel." This statement, like many popular slogans, is in one sense quite true, but in other senses quite false; in other words, it is in general quite false. If one means that hydrogen, burned as a fuel, produces pollutants in concentrations at or below current legally-allowed pollution levels, then hydrogen can be pollution-free. On the other hand, if one means that hydrogen fuel systems produce no pollutants, then much more remains to be said. The distinctions between these paraphrases often are not appreciated by the general public or by scientists inexperienced in the standards-setting process.

Contrary to common belief, legal environmental standards in the U.S. are not necessarily set at levels that eliminate environmental damage. For example, the 60-$\mu g/m^3$ air standard previously mentioned corresponds to an average visibility of only 12 to 15 mi,[1] which represents considerable environmental damage in present rural America or the Rocky Mountain West. Standards, by law as well as by economic and political reason, are constrained to levels that are achievable virtually everywhere using current, or shortly imminent, technology.

Environmental standards setters are well aware that standards are compromises and that the current technology available to meet a standard puts a practical limit on how protective a standard can be. Once the standards have been set, however, technologists tend to use them as their sole goals of environmental protection, believing there is no environmental need for trying to do better. Thus, the environmental advantages of a wholly new technology, such as hydrogen, can be lost and have difficulty breaking into the formal body of current research objectives. This greatly retards the adoption of environmental advantages into environmental law.

Standards also are limited, by and large, to dealing with existing problems caused by the pollutants emitted by existing technologies. Future large-scale technologies, such as hydrogen technology, are not covered. Thus, the statement that hydrogen meets all applicable environmental standards has no real significance in itself.

2.1.4. Significant Deterioration

Two large classes of environmental problems are frequently overlooked until after the fact. The first class is problems of new pollutants. Because no damage from a new pollutant has yet been demonstrated, it is legally difficult to establish restrictions on them. It is especially difficult in the face of the usual well-organized legal opposition to the establishment of environmental standards in general. The second class of environmental problems is problems of expanding sources. Established emission control measures for a specific activity may become inadequate as that activity expands greatly in size and territory.

Efforts have been made to develop new, legally viable concepts of environmental protection which help close these loopholes. One such development is the prohibition of significant deterioration of ambient air quality. This concept developed in the early 1970s as a result of a U.S. Supreme Court case brought by the Sierra Club and citizens' groups from San Diego, New Mexico, and Washington, D.C. The principle of "no significant deterioration" does not mean, as is sometimes stated, that no emissions are allowed. Exactly what it does mean has been the subject of several years of effort on the part of the Federal Environmental Protection Agency, the U.S. Congress, and others to find a useful and workable definition of "significant." At this time, it appears that one provision for preventing significant deterioration will be a requirement to apply the best available pollution control technology for certain prescribed pollutants. Such a requirement eventually could have powerful implications for hydrogen.

In this writer's opinion, significant deterioration of existing good air quality can be avoided by the use of the best control technologies. However, institutional barriers exist to the timely adoption of high-quality controls. Polluters are naturally hesitant, since high-quality controls as a rule cost more than rudimentary controls. Polluters are also hesitant because pollution control in many industries is a new field representing a degree of technical sophistication unusual to their long-established operations. Polluters, therefore, generally work to the best of their abilities to erect and utilize various legal and political barriers to the implementation of advanced control technology. Transcripts of numerous regulatory hearings verify this statement.[2,3]

2.1.5. New Paths

A search of the literature relating to hydrogen and its new applications reveals that little has been done to date toward quantifying total environmental impacts of large-scale usage of hydrogen. To be sure, much good effort has gone into gathering the data needed as input to such an environmental assessment. Emissions from hydrogen-fueled internal combustion engines have been measured. These emissions and factors affecting them are described in detail in Volume II of this series in the discussions of the hydrogen automobile and the hydrogen airplane.

The environmental impact, however, consists of more than the emissions per vehicle. An environmental axiom is that everything is connected to everything else. Automobile emissions apparently would be reduced very significantly by a change to hydrogen fuel, but somewhere power plant pollution must increase from producing the power to make the hydrogen to run the car. These cumulative environmental impacts have not been extensively quantified.

That they have not been quantified does not mean they cannot be quantified. Contrary to the beliefs of many people, environmental considerations are not a sort of quasi-philosophical attachment which follows after hard technical and economic considerations have been made. Nor are environmental considerations the exclusive province of the natural scientist; they call equally for the skills of the physical scientist and engineer. Environmental impact is fully as quantifiable as is economic impact (which is not to say it is perfectly calculable, either).

It has traditionally been true of new technologies, and, unhappily, it is still largely true of new hydrogen technologies, that in-depth environmental analyses come after or near the end of the other developmental work. Compare the work done on production technology or on aircraft applications in this series with the work done on environmental assessment.

This chapter seeks to point out the basic data available. It seeks to set out the directions and types of firm considerations which need to be taken in the hydrogen field, and it presents the rudimentary calculations that now exist along these directions. If this chapter serves to raise the right questions and encourages programs to obtain complete and timely answers, it will have served well indeed.

2.2. MOBILE SOURCES OF POLLUTION

The environmental advantage most commonly claimed for hydrogen is in cleaning up automobile exhausts. For this reason, we plunge into this subject first.

2.2.1. Existing Problems

2.2.1.1. Ground Transportation

There is no need to expound at length on the magnitude and implications of the environmental problem resulting from automobile exhaust, but a quick perspective will be helpful. Table 1 shows total national air pollution emissions categorized by emitting source. The contribution from transportation is obviously major. Pollution controls on autos are improving, but the number of autos is also increasing at a high rate. By about 1990, total emissions from autos are again expected to reach 1970 levels despite a fourfold improvement mandated in the effectiveness of controls (see Figure 1). Even relatively small metropolitan areas like Albuquerque, N.M., with its population of a third of a million and almost no heavy industry, suffer air pollution alerts occasionally and with increasing frequency. If automobiles really meet stricter proposed standards (Figure 1), significantly cleaner air would be enjoyed beyond the year 2000. The battle over these tougher 1976 requirements is unresolved at this writing.

Small cars and mass transit systems would help the environment, but, to a degree, they imply a change to a lifestyle which many would think un-American. Hopefully, this attitude may change with time. In any event, the automobile pollution problem is a long way from solution even if the proposed emission reductions are made, to mention nothing of the oil supply problem.

2.2.1.2. Air Transportation

Environmental problems associated with airplane emissions are more a result of where pollutants are emitted than of the sheer quantities emitted. Noise around airports constitutes a large and growing problem. Lawsuits and reduced property values are two direct, economic, and quantitative measures of environmental damage

TABLE 1

Estimated Emissions of Air Pollutants, by Weight Nationwide 1971 (Millions of Tons per Year)

Source	CO	Particulates	SO_2	HC	NO_X
Transportation	77.5	1.0	1.0	14.7	11.2
Fuel combustion in stationary sources	1.0	6.5	26.3	0.3	10.2
Industrial processes	11.4	13.6	5.1	5.6	0.2
Solid waste disposal	3.8	0.7	0.1	1.0	0.2
Miscellaneous	6.5	5.2	0.1	5.0	0.2
Total	100.2	27.0	32.6	26.6	22.0
% change 1970 to 1971	-0.5	+5.9	-2.4	-2.6	0

From U.S. Council on Environmental Quality, 4th Annual Report, U.S. Government Printing Office, Washington, D.C., 1973, 266.

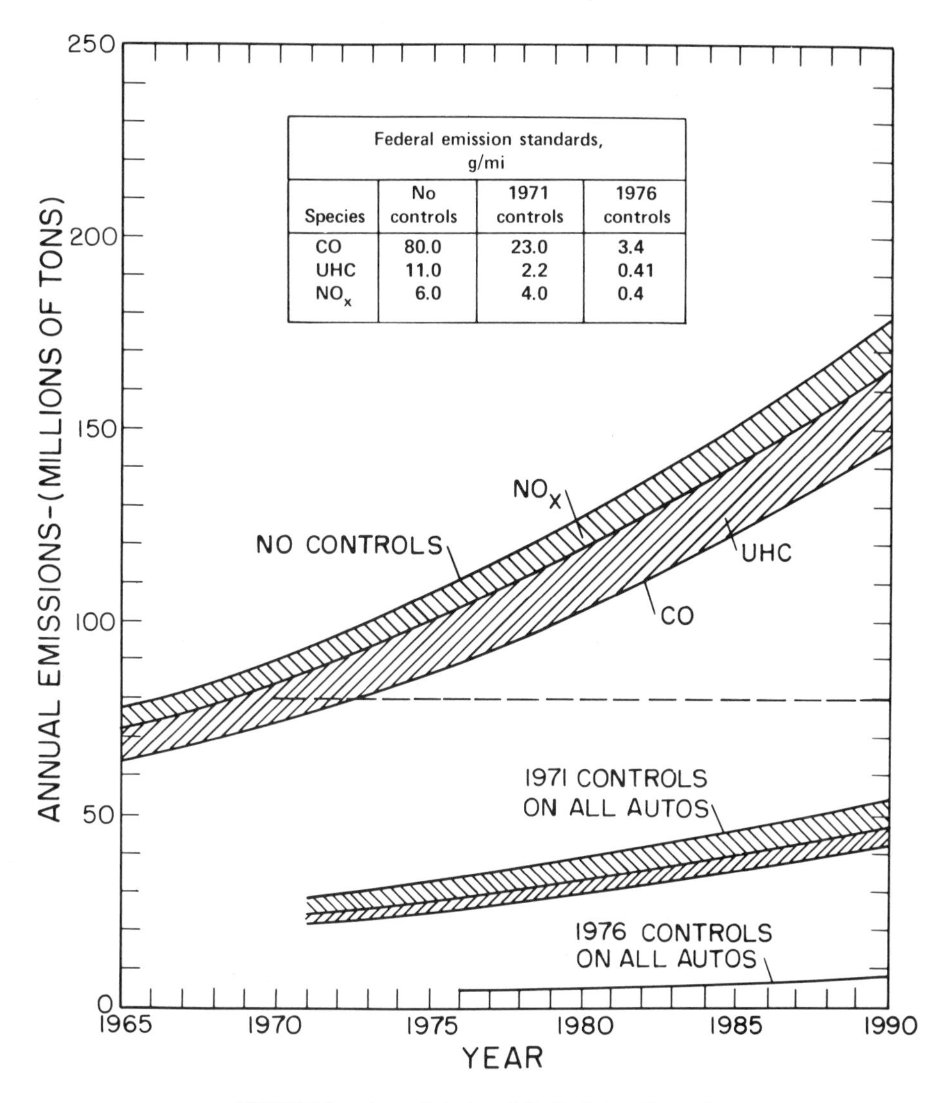

Federal emission standards, g/mi			
Species	No controls	1971 controls	1976 controls
CO	80.0	23.0	3.4
UHC	11.0	2.2	0.41
NO_x	6.0	4.0	0.4

FIGURE 1. Annual Automobile Emission Controls

from airport noise. Material pollutants – primarily unburned hydrocarbons, oxidants, sulfur oxides, and oxides of nitrogen – are of significance around airports, but, more importantly, are becoming of increased concern when they are emitted at the higher altitudes characteristic of newer and faster airplanes. Dispersive forces are small at high altitudes, and chemical reactions at these altitudes potentially have very significant and long-lasting implications for the earth's environment.

2.2.1.2.1. SSTs and the ozone layer – A classic case study of confrontation, challenge, and response of business concerns to environmental concerns can be seen in the case of the SST controversy. It is illustrative of typical moves and countermoves involving science and scientists, public information, public-relations people, and news media which typically interact to leave the man in the street confused and uncertain of somebody or other's credibility.

By 1970, a national debate was growing over the desirability of continuing national funding for development of supersonic transport, the SST. Such a plane would fly higher than conventional craft; in fact, it would fly well above 39,000 ft, which typically is the bottom of the stratosphere. This is also a region in which relatively high concentrations of ozone are found. The important role which this ozone layer plays in screening the earth from harmful ultraviolet rays is widely accepted among scientists.

Specialists in the chemistry of nitrogen compounds came to realize that SSTs would exhaust nitrogen oxides, NO_X, into this ozone layer. They realized that such compounds would catalyze free radical reactions, resulting in the partial destruction of ozone. Calculations of the quantities of pollutants emitted from assumed commercial numbers of SSTs confirmed a potentially serious problem.

At this point, aircraft companies put their scientists to work and soon publicized calculations showing no potential ozone problem from SSTs – exactly the opposite of the earlier conclusion. The reenactment of this scene – one environmental conclusion from one set of scientists and a seemingly opposite one from scientists working on behalf of an industry – has become almost *de rigueur* in environmental issues. Why? Is science really so impotent in quantifying effects? No, the answer lies elsewhere.

To settle the SST issue, the U.S. Department of Transportation (DOT) sponsored a 3-year, $20 million study whose results were released December 4, 1974.[7] The results appear to be sound and agreeable to scientists on all sides of the issue. Notwithstanding this unanimity, however, the lay public was told by each side (environmental sympathizers and industrial sympathizers) that the DOT report substantiates their original conclusions (i.e., that there is a potential ozone-layer problem and that there is no potential ozone-layer problem, respectively). The lay public is still confused.

What the DOT report actually says is that if full commercial fleets of SSTs fly using today's engines, a problem will result. For smaller numbers, and certainly for the first SST, or if engine (or fuel) modifications are made to reduce NO_X, no problem will result. The public is confused because the bases for calculations and the assumptions made are not publicized with the conclusion.

Science and scientists must do better in this regard if they are to enjoy the continued trust of the public and public policy makers. Scientists must strive to ensure that their findings are used to inform and educate the public and must not allow findings to be presented for special purposes in a perhaps true, but misleading, light.

2.2.1.2.2. Land use – The satisfactory siting of new and larger airports will become increasingly troublesome in the future as the U.S. approaches complete utilization of all lands. Recognition of this problem resulted in the passage of the Federal Airport and Airway Development Act of 1970. This Act sets out requirements for public participation to consider the "economic, social, and environmental effects of the airport location and its consistency with the goals and objectives of such urban planning as has been carried out by the community."[5]

In addition to problems of undesirable noise levels, odors, and air pollutants around airports, the elementary question of land use arises. Modern airports are large consumers of land, often of productive farmland, which causes land-use conflicts. In Japan, where land-use conflicts are further advanced than in the U.S., one consideration in favor of developing a high-speed train to run from Tokyo to Osaka instead of an airplane was the consideration that the train stations and right-of-way used up less land than two airports.

As will be seen, the application of hydrogen fuel to aircraft has favorable potential impacts on each of the environmental problems just discussed: chemical emissions, noise, and land use.

2.2.2. Impact of Hydrogen on Existing Problems

2.2.2.1. Ground Transportation

It is axiomatic that, except for possible lubricating oil consumption, an automobile running on hydrogen cannot produce any unburned hydrocarbons, carbon monoxide, sulfur oxides, or particulates in its exhaust. Using air as the oxidizer, it can, of course, produce and emit oxides of nitrogen, the fifth of the five common air pollutants.

NO_X emissions from hydrogen-fueled automotive internal combustion engines and their dependency on operating variables are discussed in detail in Chapter 1 of Volume IV of this series. The general conclusions are that NO_X emissions probably can be cut to well under the 1976-77 federally allowable standard of 0.4 g/mi, perhaps as low as 0.04 g/mi or one tenth of this standard.[4] For hydrogen-oxygen fueled cars, such as the Perris Smogless Auto, all NO_X emissions are, of course, eliminated.

What would a switch to air-breathing, hydrogen-fueled cars mean for the environment? The rate and extent to which present auto smog would disappear from urban areas is a complicated function of the rate of introduction of new cars, the rate of scrapping of old cars, the growth in numbers of cars, pollutant emission levels from new cars, and effect of age and maintenance on emission levels of new cars. These environmental studies have not been performed for a hydrogen car, and many of the basic data have not been obtained.

Of the many studies of hydrogen auto emissions, only one has reported finding any significant new air pollutant in exhaust gas. Griffith[6] analyzed exhaust gases from a Briggs and Stratton engine run on hydrogen and found hydrogen peroxide in concentrations of 220 ppm. For comparison, NO_x emission goals to meet 1976 federal standards correspond approximately to 40 ppm.

If these emissions of H_2O_2 are borne out to be typical, a potentially serious new environmental problem clearly exists. Griffith[6] suggests catalytic converters as a solution. If this proves to be the case, one hopes that the solution will be implemented from the first instead of reasoning that the environment can tolerate the H_2O_2 emissions from the first cars. Technology no longer need wait until an environmental problem already exists before acting to head it off.

2.2.2.2. Air Transportation

The replacement of aviation fuel by hydrogen is generally conceded to have the best chance of being the first wide-scale civilian application of hydrogen. This eventuality would have a favorable impact on each of the three largest environmental problems of air transportation: air pollution, noise, and land use.

Noise levels for a hydrogen-fueled airplane would be reduced below the levels expected from a hydrocarbon-fueled craft designed for a similar mission. Reductions can be realized owing to two inherent properties of H_2: low density and high heat of combustion per pound. The lower fuel density permits total plane weight for H_2-fueled craft to be about two thirds that for a comparable HC-fueled craft. Lower plane weights mean more favorable plane configurations, higher flights, reduced power, and/or steeper ascents. A higher heat of combustion means that less air per unit time must rush through an engine for a given thrust. All of these factors contribute to a quieter environment. These points are discussed in greater technical depth in Chapter 2 of Volume IV of this series.

Again, as with other applications, the direction of the environmental changes latent in hydrogen is known, but quantification of these changes in the aggregate has not been performed. In the final analysis, it is this concern with the aggregate, with cumulative effects, that marks the border between genuine concern for the environment and the traditional concerns of "can the camel carry a pound of straw?"

The property of significantly reduced weight of hydrogen-fueled aircraft has the potential to reduce land requirements (i.e., runway lengths) for airports. No calculations have been made to determine if this could have a significant impact.

The question of air pollution from hydrogen aircraft is more mixed than the noise question. Particulates, sulfur, hydrocarbons, CO, and CO_2 would be virtually eliminated. NO_x could be reduced. Water vapor, however, not normally considered to be a pollutant, would be increased by two to three times. Water vapor deposited high in the stratosphere may have detrimental effects not found when water is deposited in the troposphere. Water vapor was considered in the final report of the Climatic Impact Assessment Program (CIAP) sponsored by the U.S. Department of Transportation.[7] The report concluded that water vapor was about half as worrisome as SO_2 for present fuels. The effect of both is to increase the optical thickness of the earth's atmosphere, thereby decreasing the amount of sunlight falling on the earth and in turn decreasing its mean temperature. Mean temperature affects crops, winds, and rainfall in complex ways.

If hydrogen were to become the future fuel, the potential sulfur problem would be eliminated, but the cooling effect of the water vapor would exceed the presently projected cooling problem from the sulfur. Furthermore, short of condensing the exhaust and returning it to the troposphere, there appears to be no way of controlling the water problem other than controlling the total numbers of planes that fly. The number of hydrogen-fueled SSTs that would not exceed any given level of effect was, of course, not addressed in the CIAP report. However, very rough extrapolations would suggest small cause for concern until fleets exceeded at least several dozen, and perhaps considerably more. Clearly, more work is needed on the question of water vapor, including close monitoring of effects as the first SSTs go into operation.

2.3. STATIONARY SOURCES OF POLLUTION

"There is no free lunch" has become almost a watchword in this age of environmental awareness, and not without good cause. The environmental price tag on the environmental benefits that accrue from using hydrogen as a fuel is the disbenefits associated with expanded primary energy production.

Chemically free hydrogen is not found in nature, as are natural gas and oil. It must be separated from compounds, such as water or hydrocarbons, in which it is bound. This separation requires energy. The methods of producing hydrogen from its compounds is the subject of Volume 1 of this series. Our purpose here is to explore the effects on the environment resulting from widely increased reliance on hydrogen as a portable fuel.

2.3.1. Primary Energy Sources

The number and size of power plants in the U.S. will increase for some time to come. The increase will more or less depend on such things as population growth and whether or not energy conservation comes to be taken seriously. In any event, a hydrogen economy implies a corresponding increase in primary energy sources. There are limits to the ultimate expansion of the primary energy industry. In oversimplified terms, but succinctly and eloquently, cartoonist Bill Mauldin makes the point (Figure 2).

The first question to be answered is how much the primary energy supply must be augmented to support a full-scale hydrogen economy. This question was addressed by the Synthetic Fuels Panel convened in 1972 by the Federal Council on Science and Technology and the U.S. Atomic Energy Commission.[8] The panel concluded that "just to meet one-half of the projected transportation fuel needs for year 2000 with electrolytically produced hydrogen would require an additional

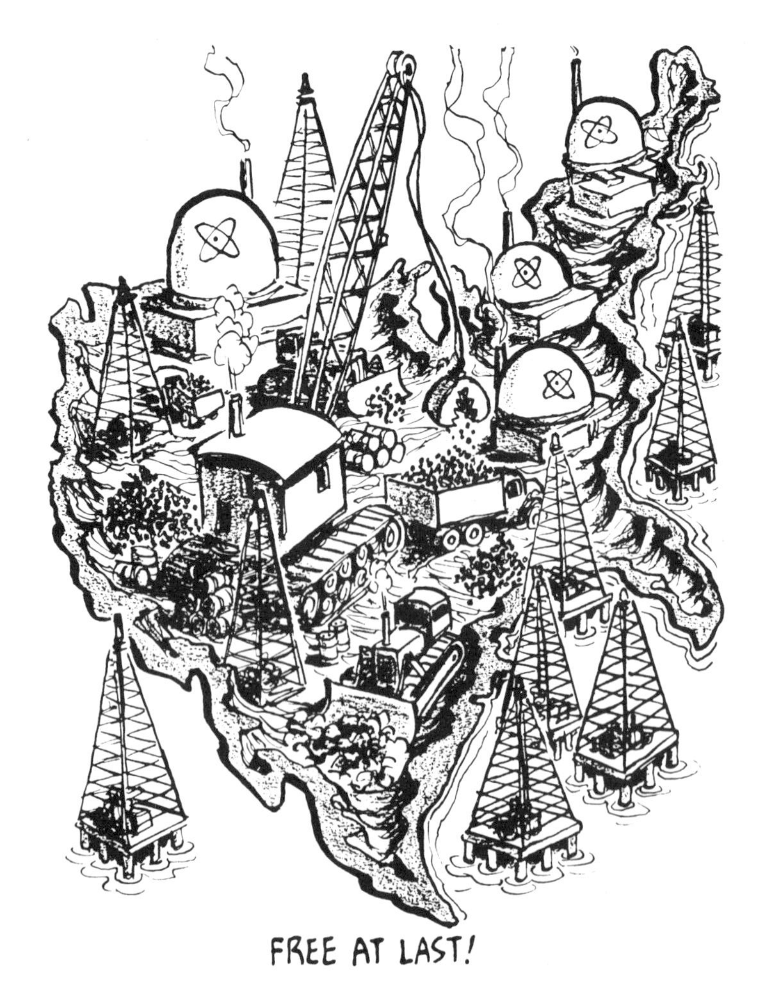

FIGURE 2. Copyright© 1975 the Chicago Sun-Times. Reproduced by courtesy of Will-Jo Associates, Inc. and Bill Mauldin.

electrical generating capacity of nearly 1,000,000 MW or about 2½ times the currently expected nuclear generating capacity at that time." To put 1,000,000 MW into another perspective, the present total installed generating capacity of the U.S. is about 500,000 MW.[9]

Gregory and co-workers[10] at the Institute of Gas Technology have addressed the question of augmented electrical capacity for hydrogen production from a slightly different tack. They examined the electrical requirements (based on three different H_2 production methods) for supplying enough hydrogen to (1) replace the actual 1968 natural gas consumption, (2) replace the projected natural gas consumption of the year 2000, or (3) replace all fossil fuels, other than those used for electricity generation, by the year 2000. Their results are shown in Figure 3. Again, we see that 1 to 5 million MW added capacity will be needed – or 2 to 10 times the present total U.S. capacity must be added for hydrogen production, beyond the capacity needed to cover other increases in electrical demand.

This energy is required just for the production of hydrogen. If it is found advantageous to ship or store hydrogen for transportation use in liquid form, considerable additional energy will be needed to cool and liquefy hydrogen gas. This would amount to approximately 250,000 to 1,000,000 MW additional.[11] Ideas have been put forward for making use of this energy of refrigeration in such ways as space cooling and refrigeration, supercharging auto engines, driving heat engines, and enriching the oxygen content of combustion air.[12] The extent to which any of these proposals might prove feasible is unknown at this time, but a potential and need exist for the imaginative exercise of energy conservation in this area.

Once the relative increase needed in energy production to supply hydrogen has been assessed, the next step is to assess the environmental impact associated with this energy production. The general subject of environmental degradation and energy production is almost a new scientific speciality in itself and is outside the scope of this

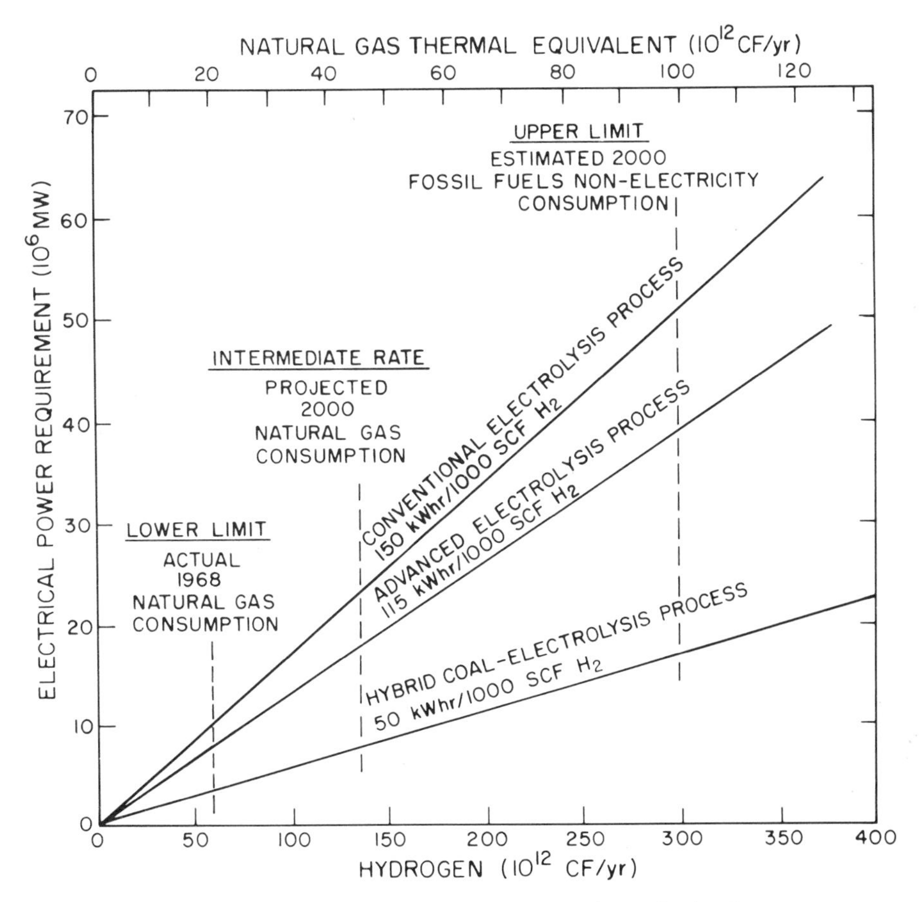

FIGURE 3. Electricity requirements for Hydrogen Production.

book. A number of treatises are available on the subject.[13-15] A partial listing of environmental issues related to energy production will recall to mind the range of problems involved: strip mining, air pollution (particulates, SO_2, sulfates, NO_x, trace metals), oil spills, thermal emissions, nuclear waste storage, emergency core cooling, uranium mill tailings, dams, etc. All of these problems would be aggravated by the need to manufacture hydrogen. It must be stated that this situation is not unique to hydrogen. The manufacture of any other synthetic fuel would similarly add to national energy supply problems.

The linking of primary energy sources to hydrogen production plants introduces operating characteristics different from today's energy applications. Some of these characteristics will affect the environmental impact.

Efforts will surely be made to generate hydrogen using off-peak power. This would improve the utilization of power plant equipment and minimize capital investment. However, it also means more plants would be run closer to full capacity during the night. Air dispersion typically is poorer at night than during the day, when insolation increases mixing. Thus, the negative impact on air quality could be proportionately worse than would be predicted by a straight-line extrapolation based on total tons emitted.

Furthermore, NO_x emissions from a fossil fuel-burning plant increase disproportionately with load. That is, emissions increase faster than output as output approaches, or exceeds, 100% of rated capacity. With higher overall plant utilizations, NO_x emissions would be expected to increase by more than the simple increase in total power produced. Because of the relatively poor state of control technology (compared to particulate and SO_2 control technology), NO_x is already the pollutant which first limits additional plant expansion in many areas of the country. Improved technology or adaptation of Japanese NO_x control technology could alter this situation.

2.3.1.1. Net Effect on Emissions

Very little has been done toward quantifying the net effect of the decreased pollutant tonnage from the transportation sector and the increased pollutant tonnage from power production in a hydrogen economy. The net effect depends, of course, on the unknown mix of supply sources for energy used in the indefinite future. A significant solar input would add virtually no air pollutants, whereas, in the near future, the predominately coal or nuclear sources needed would add significant amounts of various pollutants.

The 1972 Federal Synthetic Fuels Panel[8] completed a few sample system analyses to examine the overall net effects of replacing portions of the U.S. transportation fuel requirements with hydrogen in the year 2000. Five different cases were evaluated for resource consumption, annual costs, and emissions wastes. Various sets of assumptions were made as to possible energy schemes; for example, using all of the off-peak power to generate hydrogen to satisfy 50% of the automotive requirements, 50% of the aircraft requirements, or all of the diesel fuel requirements; or using all-nuclear off-peak power to do the same thing; or using coal to produce hydrogen through gasification. Figure 4 shows the results from one such case, the all-nuclear case above.

As expected, one sees decreased chemical emissions from the transportation sector but increased radioactive emissions and wastes from the stationary source sector. The increased radioactivity exposure is not necessarily large in absolute terms, being less than 5% of the natural background if averaged over the entire population.

It is interesting that the Synthetic Fuels Panel found *increased* total chemical emissions for the case of generating hydrogen from off-peak power from the actual mix of fuels expected in the year 2000. Even though the total pollution was increased, the Panel suggested that benefits accrued from displacing emissions from urban areas to central power stations. Although this would certainly minimize the health impact, in the long run something better than pollution relocation must be done. Large central power stations offer the opportunity for applying better antipollution controls at lower total cost than individual controls. However, the city dweller must still be willing to pay for cleanup and must not lose interest as soon as the problem is shifted into some other backyard.

Again, should events conspire to make commercial hydrogen production from solar energy practicable in the future, essentially all emissions and waste products would vanish, leaving considerations of land use as the sole remaining environmental impact.

2.3.2. Environmental Impacts of Hydrogen Production Plants

The possibility of significant environmental

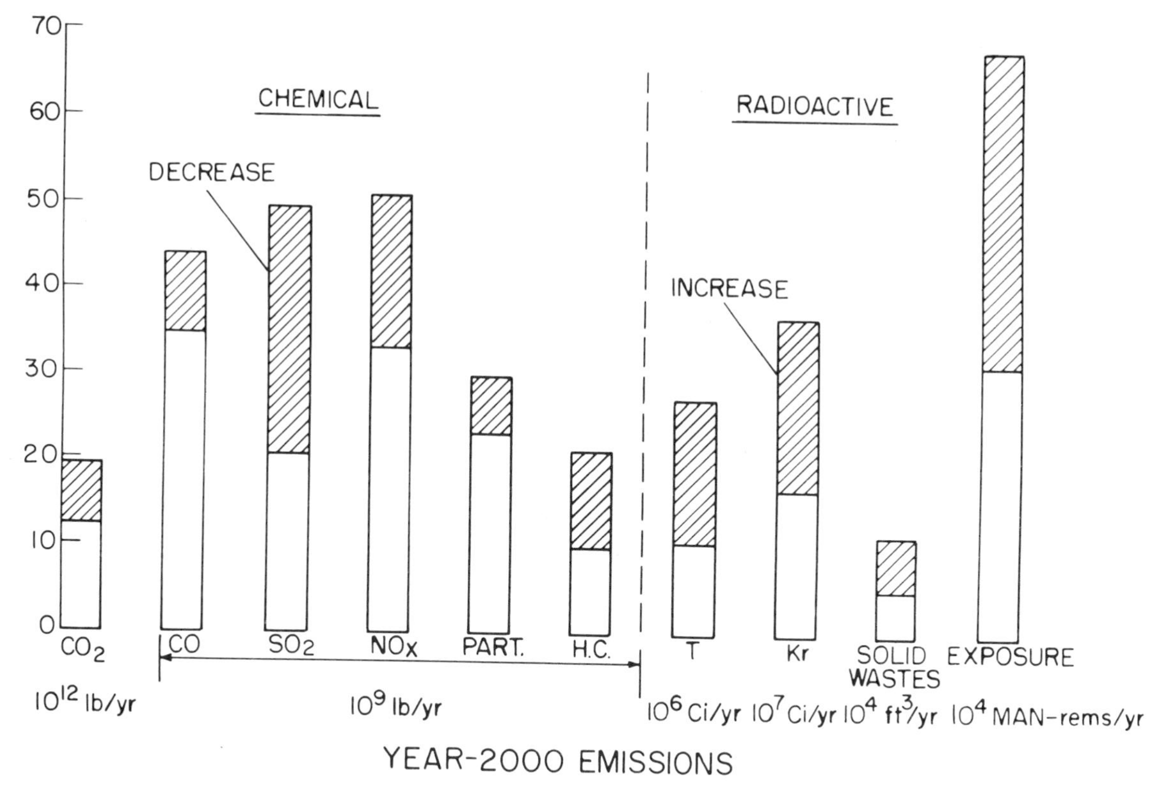

FIGURE 4. Nuclear-electrolytic H_2 supplying 50% of aircraft fuel in an all-nuclear system (adding 648 GW_e or +44%) (potential foreign exchange savings of $8 billion/year).

impacts from hydrogen plants themselves, as separate from primary energy plants, must not be forgotten. Again, little work has been done in this area. As the type of plant, much less the specific process, can only be surmised, there is little point in conjecturing emissions. Nevertheless, some general statements are appropriate.

If hydrogen were to be produced by means of the gasification of coal, the plant would be similar to plants presently proposed for gasifying coal to methane. Environmental Impact Statements (EIS) have been written for such plants.[16] The types of impacts identified as being of concern include strip mine reclamation, water availability, trace element disposition, boom town problems, and cumulative impacts on air quality, especially concerning short-term concentration peaks and effects of high terrain on air dispersion models.

A thermochemical water decomposition plant would be a new kind of plant requiring totally new analyses. Certainly the potential exists for environmental problems arising from the release of wastes from very large plants using some of the substances mentioned for thermochemical hydrogen cycles, e.g., chlorine, bromine, iodine, vanadium, sulfur, and mercury.

Water electrolysis plants are presently unregulated in this country, since there are none of sufficient size to be of any environmental concern.[17] Environmental impacts of large electrolysis plants need to be studied. There may be potential water pollution problems arising from the impurities found in the water used or from the alkali used in the solution to be electrolyzed.

2.3.3. Hydrogen Fuel Cells

Hydrogen-powered fuel cells very probably will play some role in a hydrogen economy. Two specific applications have been frequently suggested. The first is for home electricity generation, and the second is for utility peak shaving.[18,19] Environmentally, the hydrogen fuel cell is attractive because it avoids NO_x emission problems associated with combustion processes. This could have significant implications for reducing emissions in urban areas, although it is unlikely that fuel cells could be used to reduce total NO_x emissions nationwide. Fuel cells are also quiet.

As a utility energy storage device, hydrogen fuel cells potentially offer an environmentally better alternative to pumped storage in locations where the latter impairs important recreational assets. Perhaps one day fuel cells will obviate land-use conflicts like the ten-year-long legal and political struggle over Consolidated Edison's Storm King reservoir at Cornwall, N.Y.[20]

2.3.4. Emissions from Home Hydrogen Appliances

Emissions have been measured from standard and modified gas appliances burning hydrogen.[21] NO_x levels were reported as follows:

standard Al range burner (no catalyst), 150 ppm
standard Al range burner (catalyst), 5.25 ppm
stainless steel experimental burner (catalyst), 1.8 ppm
Al oven/broiler burner (catalyst), 4.5 ppm
cast iron space heater (catalyst), 4 ppm

2.4. OTHER ASSOCIATED ENVIRONMENTAL EFFECTS

The implementation of a large-scale usage of hydrogen in the manner discussed in these four volumes carries with it implications for environmental changes beyond the hydrogen itself. Alternatives, by-products, and resources associated with the implementation have their own further consequences.

2.4.1. Oil Reduction

A reduced traffic in crude oil will with virtual certainty be a cause, not a result, of large-scale hydrogen usage. Nevertheless, reduced dependence on oil will be characteristic of the hydrogen economy. Numerous environmental problems today result from the procuring, handling, and transport of petroleum.

The vast majority of oil goes where it is wanted without mischief or mishap. Even so, the quantities handled have become so huge that a relatively few accidents and numerous minor discharges can, and have, caused extensive damage. In 1971, there were 8496 polluting spills to U.S. waters reported, involving about 9 million gallons of spillage.[22] Many of the incidents have been well publicized: the *Torrey Canyon* oil tanker collision in 1967, the *Ocean Eagle* in 1968, the Santa Barbara Channel oil leak in 1969, the Alaska Pipeline controversy of 1970 to 1973, and multiple blowouts and fines associated with improperly guarded oil well operations off the Gulf Coast.

A burgeoning environmental issue – although a seemingly anachronistic one in the era of a U.S. policy toward energy independence – is the issue of superports. Supertankers, carrying 250,000 tons of foreign crude, reduce the cost of shipping crude, but increase the risks of superaccidents. Superports (accommodating 90-ft drafts) are needed where today's deepest ports are 40 to 50 ft.[23] The adverse environmental impacts of superports have been studied and found sizable, but are judged by some to be less than the benefit of the oil.[24]

The disposal of salt water produced in association with crude oil has caused problems where it has been done carelessly to save costs. The volume of this brine can exceed the volume of oil by several times.

The ultimate adverse impact which oil may have on our environment is war over the sources of supply. All of these impacts will be reduced proportionally as the world's use of and dependence on oil are reduced.

2.4.2. Electric Power Transmission Lines

In 1971, aboveground power transmission lines in the U.S. occupied about 4 million acres, a larger area than the state of Connecticut. This use is expected to increase to 7 million acres by 1990.[14] As open land grows more scarce and more costly and the blocks of power grow larger, utilities have gone to higher and higher voltages for transmission. This decreases the number of lines needed, but increases the amount of land needed per line. It also raises as yet unresolved questions about the effects of magnetic and electric fields on human beings and of coronal discharges on atmospheric constituents. Transmission line corridors cause increased runoff and siltation of streams and generally imply increased use of herbicides.[25] In addition, in this writer's opinion, transmission lines are ugly.

Whether the increased use of hydrogen as an energy carrier would result in fewer or more electric power lines is an open question. With good environmental planning or favorable progress in nonelectrolytic hydrogen production methods (see Volume I of this series), hydrogen has the potential to slow or eliminate the need for further land

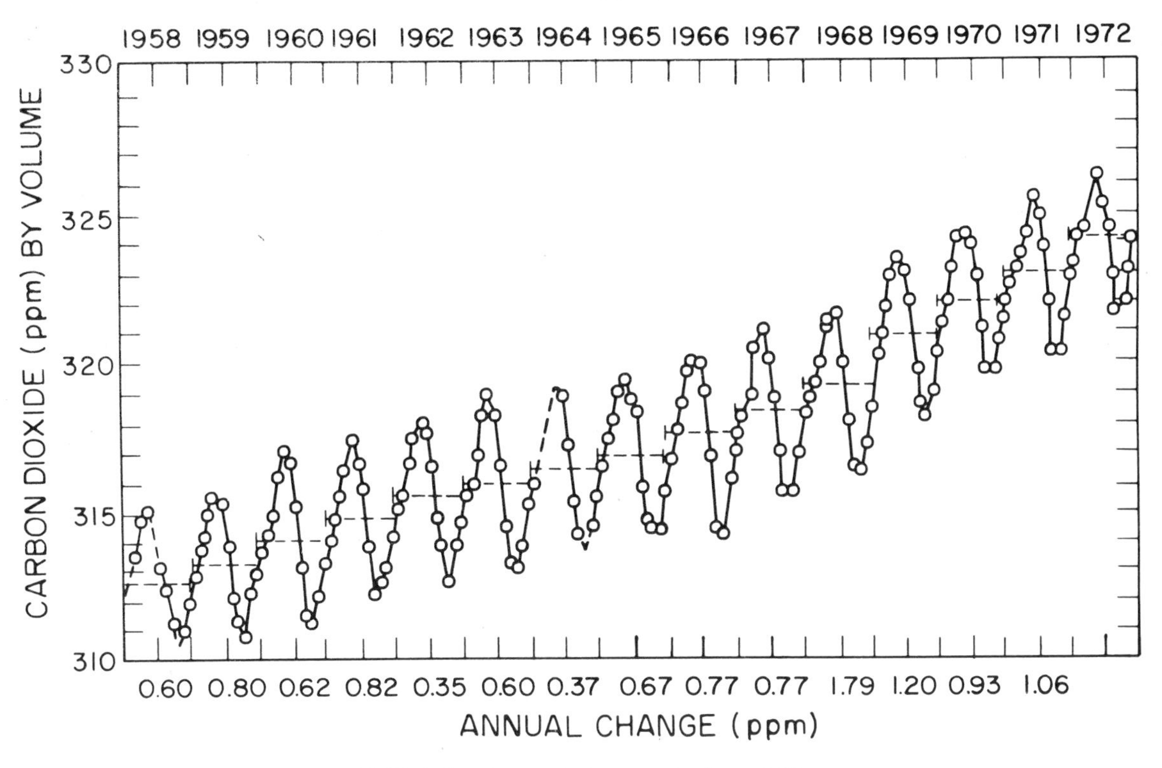

FIGURE 5. Atmospheric concentrations of CO_2.

allocation exclusively to energy corridors. Schemes for transporting energy into cities in the form of hydrogen flowing in underground pipelines are discussed in Volume II of this series.

2.4.3. Thermal Pollution from Energy Parks

One possible way to improve the overall safety of nuclear power generation is to site a number of reactors, perhaps totaling 40,000 or 50,000 MW capacity, at one location, together with all the associated fuel processing and waste handling facilities. A disadvantage of this scheme, however, is the very large local dissipation of waste heat which results. The reject heat from such a complex is of the order of the energy contained in thunderstorms. Local weather modifications could be severe.

Nuclear parks linked with hydrogen production plants to utilize the waste heat have been suggested. In effect, much of the reject energy then appears in the form of hydrogen, which is shipped away, instead of being dumped locally. This has the potential for conserving energy and simultaneously mitigating an environmental problem.

2.4.4. Carbon Dioxide Buildup

Carbon dioxide is not legally defined as an air pollutant. At concentrations found in even the most highly polluted atmospheres, it does not directly harm organisms, materials, or aesthetics. There are no restrictions or regulations on its emission. Yet there is convincing evidence that CO_2 is the most pervasive of all anthropogenic air pollutants and one with potentially damaging effects on a global scale. It is possible that climatic effects caused by the buildup of carbon dioxide in the earth's atmosphere will ultimately become the overriding incentive to establish a hydrogen economy.

Data from the Mauna Loa Observatory in Hawaii (Figure 5) clearly show a steady trend of rising atmospheric concentrations of CO_2, apparently of global extent.[26] The causes and results of this increase are less certain than is its existence. There are persuasive arguments and evidence to indicate that the cause is the man-controlled combustion of fossil fuels.[26,27] The result of CO_2 buildup is more problematic, but it is reasonable to predict that the result will be an unacceptable increase in the average temperature of the earth's

TABLE 2

Carbon Dioxide Added to Atmosphere by Consumption of Fossil Fuels

Decade	CO_2 added (ppm/decade)	Cumulative total (ppm)	Atmospheric concentration (ppm)	Predicted temperature rise (°C)
1860–69	0.6	0.6	295	0.00
1870–79	1.0	1.6	296	0.00
1880–89	1.4	3.0	296	0.01
1890–99	2.1	5.1	297	0.02
1900–09	3.4	8.5	298	0.03
1910–19	4.6	13.1	300	0.04
1920–29	5.3	18.4	302	0.06
1930–39	5.6	24.0	305	0.08
1940–49	7.2	31.2	307	0.10
1950–59	10.3	41.5	312	0.13
1960–69	14.5	56.0	317	0.18
1970–79	19.9	75.9	325	0.24
1980–89	28.0	104.0	336	0.33
1990–99	38.5	142.0	352	0.46
2000–09	53.4	196.0	373	0.63
2010–19	78.0	274.0	405	0.88
2020–29	110.0	384.0	449	1.2
2030–39	154.0	538.0	510	1.7
2040–49	216.0	754.0	597	2.4
2050–59	304.0	1058.0	718	3.4
2060–69	428.0	1486.0	889	4.8
2070–79	602.0	2088.0	1130	6.7

surface.[27] Table 2 shows this predicted increase.

The increase is due to the strong absorption of radiant energy in the infrared region exhibited by CO_2. Since more of the radiant energy leaving the earth than coming to the earth is in the infrared, an increase in CO_2 will trap energy on the earth and cause its temperature to increase. The effect that a 2°C rise in average temperature would have on weather, crops, ice caps, ocean levels, and CO_2 release from ocean waters is still more uncertain. However, energy systems based on hydrogen in water instead of the carbon in fossil fuels would, of course, bypass the CO_2 cycle entirely.

2.4.5. Oxygen By-product

Oxygen is a by-product of all hydrogen production processes involving thermochemical or electrolytic splitting of water. This tonnage of oxygen, eight times the tonnage of hydrogen produced, could probably be dispersed to the atmosphere safely, although apparently no studies have been made of possible localized effects from giant hydrogen plants.

Numerous possibilities for environmental benefits from this by-product oxygen can be easily conceived. Several large classes of environmental problems are caused or aggravated by either an oxygen deficiency or the impurity of oxygen in combustion air. A ready source of cheap and relatively pure bulk oxygen would be an environmental bonanza if economics permitted its distribution and use.

2.4.5.1. Water Pollution and Oxygen

Among the fundamental measures of water pollution are three parameters related to the oxygen balance: Biological Oxygen Demand (BOD), Chemical Oxygen Demand (COD), and Dissolved Oxygen (DO). Reduction or exhaustion of DO is one of the most harmful effects of many water pollutants – including many industrial chemicals, municipal sewage, agricultural runoff (fertilizers), and feedlot wastes. Oxygenation of these waters, either before or after they entered tributaries, could in the long run prove to be the most economical antipollution treatment.

2.4.5.2. Air – 79% Diluent

The nitrogen associated with oxygen in air contributes to a number of the environmental problems characteristic of today's industrialized nations. Oxides of nitrogen are formed at the temperatures reached in the combustion of air-fuel mixtures. Although the concentrations in exhaust gases may be only several hundred parts per million NO_x, this amounts to over 20 million tons annually. In terms of tonnage, NO_x is the fifth largest air pollutant and the fastest increasing in the U.S.

The large dilution of hot stack gases caused by large volumes of unneeded nitrogen passing through industrial combustion processes is additionally a source of much wasted energy. This nitrogen also significantly increases the cost of treating flue gases to remove sulfur dioxide and other pollutants. The use of oxygen for combustion would alleviate this whole class of environmental problems.

2.4.6. Resource Commitments

A commitment to large-scale applications of hydrogen implies a commitment of certain natural resources as well. The National Environmental Policy Act (NEPA) of 1970 requires environmental impact statements preceding all major federal actions. The statement must include discussion of any irreversible and irretrievable commitments of resources involved in the implementation of the proposed action. The purpose obviously is to force planners to examine the impact of their decisions on the nation's or the earth's finite supply of natural resources. This examination has not yet begun in earnest for most proposed applications of hydrogen.

Some general remarks of illustrative nature are all that can be offered. Large-scale use of metal hydrides for hydrogen storage (see Volume 2 of this series) has implications for the extractive industries. If metals of limited domestic supply are found to be advantageous, there are implications for the U.S. balance of payments and national security. In the space program, large-scale use of cryogenic hydrogen called for the consumption of large amounts of helium gas for purging and pressurization. Perhaps these functions could be performed in other ways with equal safety, without expending such a unique resource as helium.[28]

The best electrode materials for hydrogen fuel cells are precious metals, such as platinum, palladium, and rhodium. Nickel can also be used. All of these are materials for which at least three quarters of the present U.S. demand is supplied by foreign countries.[29]

2.5. CONCLUSION

The conclusions to be drawn in the middle of the 1970s, as the U.S. enters the decades when synthetic fuels must begin to take up the energy loads carried by fossil fluids, are conclusions of direction rather than destination.

First, hydrogen usage has widespread ramifications and many apparent advantages for the environment. Second, the use of hydrogen alters, but does not eliminate, environmental limitations on expanded energy production.

It is hoped that this chapter will set seeds to produce early, candid, and quantified analyses of hydrogen systems and the environment. Going further, it is hoped that those studies will be used, without regard for personal niches, to optimize future energy systems, with environmental impact being among those parameters which are minimized. Toward this goal, it is hoped to dispel the impeding notion that environmental science is somehow a "soft" science not amenable to calculation. Environmental science is a "hard" science, but an immature science. The immaturity of the science is revealed in this survey of analyses which remain to be done on environmental effects.

Finally, no chapter on the environment can properly be concluded without reemphasizing the need for energy resource conservation regardless of the road taken.

REFERENCES

1. U.S. Dept. of Health, Education, and Welfare, *Air Quality Criteria for Particulate Matter,* National Air Pollution Control Administration publication No. AP-49, January 1969, 3.
2. State of New Mexico Environmental Improvement Agency, Transcripts of Air Pollution Hearings, September 1969, October 1971, August 1974.
3. Hearings before the Subcommittee on Environmental Pollution of the Committee on Public Works, U.S. Senate, 94th Congress, first session April 24, 29, 30, and May 1, 1975, Implementation of the Clean Air Act – 1975, U.S. Government Printing Office, Washington, D.C., 1975.
4. LH_2 – a more efficient fuel for cars, *Cryogenics,* 15(10), 615, 1975.
5. U.S. Council on Environmental Quality, 3rd Annual Report, U.S. Government Printing Office, Washington, D.C., 1972, 212.
6. **Griffith, E. J.,** Hydrogen fuel, *Nature,* 248, 458, 1974.
7. **Grobecker, A. J., Coroniti, S. C., and Cannon, R. H., Jr.,** The effects of stratospheric pollution by aircraft, DOT-TST-75-50, December 1974, available from National Technical Information Service, Springfield, Va.
8. U.S. Atomic Energy Commission, Hydrogen and other synthetic fuels, TID-26136, September 1972.
9. U.S. Dept. of Commerce, *Statistical Abstracts of the United States,* U.S. Government Printing Office, Washington, D.C., 1975.
10. **Gregory, D. P., Ng, D. Y. C., and Long, G. M.,** The hydrogen economy, in *Electrochemistry of Cleaner Environments,* Bockris, J. O., Ed., Plenum Press, New York, 1972, chap. 8.
11. **Edeskuty, F. J.,** Los Alamos Scientific Laboratory, private communication, 1976.
12. **Parrish, W. R.,** Recovery of hydrogen liquefaction energy, in 10th Intersociety Energy Conversion Engineering Conf. Record, The Institute of Electrical and Electronic Engineers, Newark, Del., 1975, 1352.
13. **Odum, H. T.,** *Environment, Power, and Society,* Wiley-Interscience, New York, 1972.
14. **Fabricant, N. and Hallman, R. M.,** *Toward a Rational Power Policy – Energy, Politics, and Pollution,* George Braziller, New York, 1971.

14a. **Wilson, R. and Jones, W. J.,** *Energy, Ecology, and The Environment,* Academic Press, New York, 1974.

15. **Fowler, J. M.,** *Energy and the Environment,* McGraw-Hill, New York, 1975.
16. Western Gasification Company Coal Gasification Project and Expansion of Navajo Mine by Utah International Inc., New Mexico, Environmental Statement, U.S. Dept. of the Interior – Bureau of Land Reclamation, Salt Lake City, 1975.
17. **Goad, M.,** New Mexico Environmental Improvement Agency, Water Quality Section, private communication.
18. **Gregory, D. P., Ed.,** *A Hydrogen Energy System,* Institute of Gas Technology, Chicago, 1973, VII. 24.
19. **Fernandes, R. A.,** *Hydrogen Cycle Peak Shaving,* Empire State Electrical Energy Research Corp., New York, 1975.
20. **de Pass, V. E.,** Energy vs. the environment in an urban center, 9th Intersociety Energy Conversion Engineering Conf. Proc., American Society of Mechanical Engineers, New York, 1974, 515.
21. **Baker, N. R.,** Oxides of nitrogen control techniques for appliance conversion to hydrogen fuel, 9th Intersociety Energy Conversion Engineering Conf., Proc., American Society of Mechanical Engineers, New York, 1974, 463.
22. U.S. Council on Environmental Quality, 3rd Annual Report, U.S. Government Printing Office, Washington, D.C., 1972, 118.
23. **Walters, S.,** What price the superport?, *Mech. Eng.,* 98(1), 46, 1976.
24. *Superport Studies,* Center for Wetland Resources, Louisiana State University, Baton Rouge, 1975.
25. **Kitchings, J. T., Shugart, H. H., and Story, J. D.,** Environmental impacts associated with electric transmission lines, ORNL-TM-4498, Oak Ridge National Laboratory, Oak Ridge, Tenn., March 1974.
26. U.S. Council on Environmental Quality, 4th annual report, U.S. Government Printing Office, Washington, D.C., 1973, 278.
27. **Plass, G. N.,** The influence of the combustion of fossil fuels on the climate, in *Electrochemistry of Cleaner Environments,* Bockris, J. O., Ed., Plenum Press, New York, 1972, chap. 2.
28. **Seamons, R. C.,** The Energy Related Applications of Helium, U.S. Energy Research and Development Administration, ERDA-13, 1975.
29. U.S. Bureau of Mines, U.S. Dept. of Interior, Status of the mineral industries, 1975, 6.

Chapter 3

Hydrogen Energy: Political and Social Impacts

Chapter 3

HYDROGEN ENERGY: POLITICAL AND SOCIAL IMPACTS

Jack D. Salmon

TABLE OF CONTENTS

3.1. INTRODUCTION

This chapter is concerned with the political and social impacts of large-scale hydrogen energy use, including certain political strategies that may encourage or discourage the development of a "hydrogen economy." A hydrogen economy may be defined broadly as a hydrogen-using energy system accounting for a nontrivial portion of the national energy system.

As the literature on hydrogen amply illustrates, there is no clear consensus among technologists as to what a hydrogen economy will consist of and what primary energy will power it. The preferred uses of hydrogen are also undetermined. Until some consensus on hydrogen technology and systems can be reached, sociopolitical predictions of system impact will necessarily be somewhat vague. It is possible to develop complex models, but at this point such exercises seem to be academic in the conventional sense of the term: formal, pedantic, but of little value in decision making. While the development of predictive models in social and political areas is certainly important, our presently limited skills in model building become even more precarious when based on an unknown.

Social and political prediction can be and is done daily. Virtually every policy choice, investment decision, and legislative or administrative order is premised upon a set of predictions about the future political and social environment. Many such decisions embody conscious desires to change that future from some predicted pattern to a different, preferred pattern. In this sense, anyone who proclaims the desirability of a hydrogen economy is indulging in sociopolitical forecasting.

However, social and political environments are more complex than technological environments and certainly less amenable to linear projections.[1] Short-term projections can be and usually are largely linear, simply because little short of catastrophe can alter massive social inertia within brief periods of time. About four to five years, roughly the time from the 1973 gasoline shortage to President Carter's 1977 energy program, seems to be the minimum for the political system to register major social value changes and begin to move beyond "quick fix" policies. In a medium term of 5 to 20 years, a far wider range of alternative policies can be developed. But it is important that policy choices taken do not foreclose options which may later be found increasingly attractive. It is in the short and medium terms that our still limited sociopolitical predictive capacity can be of some value, and it is also here that policy decisions exert greatest leverage.

Beyond the middle term, we enter upon problems of long-range forecasting, where neither social nor technological techniques seem capable of very reliable prediction.[2] The problem is not that long-range predictions are impossible, but that we have no way of reliably discriminating between accurate and inaccurate predictions. So many variables intervene and interact that any prediction may be true or false. Unfortunately, a hydrogen economy probably will not be realized in the middle term.

However, if individuals and societies choose to do so, they can design much of their long-term future by making middle-term choices which narrow the range of long-term alternatives which could become reality. This is essentially the point of the controversial *Limits to Growth*[3] studies: that certain highly undesirable long-term futures are indeed possible and even probable, but that by taking appropriate measures at appropriate times we may alter the probabilities toward more satisfactory futures. Those interested in a future hydrogen economy may, then, profitably seek to identify those middle-term choices which could increase the probability of a long-term hydrogen economy. In part, this means attempting to maintain a maximum number of options as long as possible in such areas as the production, transportation, and use of hydrogen, so that developments in other areas (e.g., energy crises, solar power technology, lifestyle changes) do not erode away the base of support required by a particular "locked-in" choice. Additionally, it means that one can work to obtain the "right" choices for society, a process known as politics. This makes possible the "self-fulfilling prophecy," in which a prediction of some future behavior interacts with present behavior in such a way that the prediction becomes true. An easy and appropriate example is the use by political candidates of the bandwagon approach: by exhibiting total confidence in victory and warning the undecided that they had "better not be left behind," it is sometimes possible to build sufficient support to win an election, although confidential polls indicated that victory was originally very much in doubt.

Therefore, this chapter does not offer a detailed exploration of some set of alternatives, but attempts to inventory the short-term context within which social inertia largely determines possibilities and middle-term choices which may increase the probability of a hydrogen economy.

It is increasingly recognized that any large-scale technological change or public project is virtually certain to produce "second-order consequences" and that among them will be some redistribution of social and political power, benefits, and costs.[4] As the sociopolitical counterparts of the famous (or infamous) environmental impact statement required by law since the National Environmental Protection Act of 1969, social impact assessments (SIA) are now explicitly or implicitly required in a number of areas. They will no doubt become standard components of nearly all major project evaluations in the near future.[5] Ideally, a well-done SIA will provide decision makers and the public not only with the information needed to decide whether or not to proceed with a project but with advance warning of some of the project's disadvantages and therefore provide the opportunity to take ameliorative action.

Since many of the nontechnological arguments for a hydrogen economy involve the assertion that a hydrogen system is socially preferable to various alternatives, SIA is especially appropriate. In any case, SIA will be done, either formally as hydrogen systems become sufficiently settled in design to allow systematic analysis, or informally as projects begin and prompt letters to Congressmen from worried, angry, or persuasive citizens. Although it is simply too early to do more than sketch some of the outlines of an SIA of the hydrogen economy, it is time to begin.

Because many readers of these volumes may be unfamiliar with the workings of American politics, it may be useful to introduce an analogy. The U.S. political system may be conceived as a machine designed by master engineers. As a machine for government, the federal government makes use of three subdivisions ("branches") to serve as control devices, each monitoring the the other two and acting to modify the inputs and outputs of the political process. The machine has two particular purposes: first, to govern; second, to avoid the concentration of power which characterizes dictatorial government. Further subdivision of power between the national and state governments (all of which are likewise composed of three branches with similar operating principles) adds to the control system, if not always to the output. This machinery was designed to accomplish its purpose of governing, but to be inefficient in terms of speed and input-output ratio, because efficiency was less desired than authoritarian government was feared. Thus, civil liberty is protected by a political machine which makes a virtue of inefficiency, invites confusion, and frequently produces outputs which are less than optimal if judged by some absolute standard. Good ideas can be lost in the machine; conversely, bad ideas may make it through. We are still conducting a running SIA of this piece of political technology, but it is essential to recognize that the eventual disposition of hydrogen economy schemes depends very heavily on such essentially (at this time) unquantifiable factors as luck, timing, personal foibles, and a host of others.

3.1.1. A Comment on Limitations

Three limitations govern this chapter. First, no attempt is made to evaluate the total set of possibilities which such a term as "hydrogen economy" might encompass. There are perhaps four major production methods, a dozen candidates for storage and transmission systems, and hundreds or thousands of potential uses. Until such time as some specific systems can be proposed, any SIA must either become so vague as to be of marginal usefulness or must focus on selected important categories rather than on total systems.

Second, rather than attempting to forecast, the approach of "alternative futures" seems appropriate.[6] A forecast attempts to predict a specific future which has a high probability of occurrence; alternative futures attempt to plot several possible outcomes or "scenarios," all of which are possible, but without assigning specific probabilities to any of them. Of course, some futures seem more probable than others, but attempting to forecast in detail at this stage of the game would be presumptuous. The scenario method should help to clarify some problems of choice.

Finally, this chapter will follow the apparent custom in the hydrogen energy literature and leave the problem of SIA or environmental impact statement (EIS) on the primary supply of hydrogen to be handled in work focused on specific primary sources. Surely there will be a difference if hydrogen is to be made from strip-mined coal rather than from a solar-water source, but to discuss specific processes would lead us into areas in which much work already has been and is being done, and therefore need not be duplicated here.

Very little study has been devoted to the social and political consequences of hydrogen energy, as opposed to the very considerable development of the technology. There are frequent references in the literature to social and political factors, but most are essentially variations on two themes: first, that hydrogen is nonpolluting and renewable and therefore highly desirable, but second, that it is unfairly discriminated against because of its association with exploding dirigibles. Some very recent work has gone considerably beyond this stage, but we are still very close to the frontier.[7]

Social scientists recognize, with both frustration and occasional sneers at mere engineers, that social systems are inherently far more complex and intractable than most "technical" problems. Administrator Seamans of the Energy Research and Development Administration (ERDA) has observed[8] that ". . . in my view, in the long-range the non-technical problems [of energy] will likely turn out to be more difficult to solve than the admittedly complex technical problems," a statement with which a political scientist can only agree. However, if we can usefully suggest some parameters within which the technology will probably be forced to operate, we may have taken a useful step toward system definition.

3.2. ENERGY AS A FACTOR IN SOCIETY

No one disputes the proposition that the ability to produce and control a large quantity of inanimate energy is basic to the development of a high standard of living, to industrial civilization, and to what we think of today as "modern" lifestyles.[9] Indices of "national power" have been constructed by scholars of international relations, with gross and per capita energy figures playing a prominent role as indicators of current economic and military power and of potential power. However, it is possible for two industrial societies to have quite similar living standards although the per capita energy consumption is much less in one than in the other. There is evidence that as industrial societies pass into the "postindustrial" stage in which service industry begins to account for as much or more of the GNP than do the classic heavy industries, energy consumption curves flatten. It is not clear whether this means simply that consumption increases at a decreasing rate of increase, or whether it is possible to attain a "zero energy growth" status.[10] There is also growing acceptance of the basic premise involved in the "limits to growth" studies: an infinite increase of anything, including energy production and use, is not possible in a finite world, and, therefore, some optimal energy level must exist for both the world and any given society. The location of that level is still very much disputed.[11]

3.2.1. Distribution

Distribution of energy resources is another critical social and political factor. The OPEC oil embargo and price increases have forced consideration of the importance of international distribution of energy resources. Domestically, the petroleum-poor New England states and the petroleum-rich Southwestern states have engaged in political maneuvering over tariffs, prices, and allocations. The issue of the distribution of access to resources has given rise to public and private discussion of the U.S. use of military power to appropriate Persian Gulf oil areas,[12] and to the fabled Texas bumper sticker symbolizing the ultimate social solution for resource-poor areas: "Let the bastards freeze to death in the dark." Geographic location, access, control, and ease of transportation will be significant political factors in all future energy systems, perhaps overriding economics.

The problem of the social distribution of energy has been and will be equally important. Given the fact that both internationally and nationally there are considerable differences in energy consumption between rich and poor,[13] the allocation of either additional or diminished energy output will require political decisions (remembering that a "nondecision" may have the same practical effect as a decision). These are classically difficult political questions.

3.3 ENERGY AS A FACTOR IN POLITICS

Since at least the time of Adam Smith's *Wealth of Nations* (published in 1776), it has been a commonplace that a growing economy is politically easier to live with than a static or shrinking economy. By analogy, a pie which is static in size but which must feed an increasing number of diners will soon be reduced to mere slivers of sustenance and will become a source of dispute as all seek larger pieces than their neighbors. But a pie which has the growth capacity of the Biblical loaves and fishes obviates at least the worst of the competitive struggle; even if some get more than others, all can have enough and perhaps even a share which grows, absolutely if not relatively.

Barring cataclysm, both the U.S. and world populations will continue to grow for at least the next 50 years. The "third world" will seek to develop industrially to at least a pale shadow of U.S. or European affluence, and the American under class will press for more of the pie. The problem of the distribution of energy resources, even if means are found to produce more and more energy, will remain at least as large a political issue as it is today. If growth in energy supplies is constrained to lower or static levels, the political frictions will grow apace. Since it is in the nature of political society that a need felt by large numbers of citizens will be responded to in some manner by the political institutions of the society, we may expect governments to play increasingly important roles in the energy system. It is a truism that the U.S. has had no overall energy policy, but rather many particularistic energy policies derived from the impact of vigorous private interests upon largely inattentive governmental institutions.[14] It is as clear as anything ever is in politics that this situation is changing; toward what it is changing is not yet clear. The Ford Foundation Energy Policy Report[15] provides a sample list of political outcomes. We shall return to this point below. It is equally clear that international relations will be heavily influenced by energy politics – indeed, the impact is already being felt. The International Energy Agency, founded in November 1974, has an express mission of developing methods to deal with international energy rationing in times of shortages.

3.3.1. Social and Political Values

For the U.S. in particular, social and political values peculiar to our society have affected past attitudes and policies toward energy, and will continue to do so for some time. Our expectations have always been those of a future-oriented, progress-minded society in which tomorrow will be better than today. Much of this social myth has come to be symbolized in our material wealth: the fabulous "American standard of living" which promised chickens in every pot and two cars in every garage. Although both the accuracy and the wisdom of this set of social myths and practices have lately come under considerable criticism, they continue to exert very strong force on national politics and "man-in-the-street" thinking. One need only observe the great caution with which national political figures discuss stabilizing or reducing energy consumption, and the ease with which they turn to programs to increase energy production as the preferred answer. This preference for seeking ways to maintain the social inertia of a growing energy system, in the face of the potential threat of petroleum import reductions or simple resource exhaustion, finds expression in such recent supply-oriented legislation as Public Law 94-473, the Solar Energy Research, Development, and Demonstration Act of 1974, and a number of Congressional or Presidential proposals for programs to expand production of other energy forms. It may be expected to affect hydrogen energy prospects as well. We have come to accept high living standards and abundant energy as a right, even a necessity. It is unlikely that a promising method of preserving that condition will be allowed to languish.

3.3.2. The Interplay of Interested Groups

Existing and potential energy systems must further reckon with a political fact which has been well established by hundreds of studies: policy, whether local or national, tends to be the outcome of pressures and vacuums created by the interplay of interested groups.[16] Indeed, our national political system was explicitly engineered to "check and balance" interest groups, producing a synthesized policy output.[17] In national energy policy, we accept without even thinking about it that there will be interest-group presentations, public propaganda, arguments and counterarguments, and maneuvers. Any evaluation of the potential role of hydrogen in the energy system must consider how interest-group conflict will affect or be affected by hydrogen. Figure 1 presents a diagram of the processes involved in the making of hydrogen policy.

If hydrogen is seen as a dangerous rival by the current major interests in energy policy (oil, natural gas, coal, nuclear), we may expect to discover the erection of numerous barriers, both obvious and subtle, to anything which might promote hydrogen energy. The current vested interests, entrenched behind a powerful defense of policy inertia, bureaucratic vested interest, stockholder influence, ample money for "public education," political leaders with strong ties to an interest ("oil Senators"), and other fortifications, may make the implementation of hydrogen energy virtually impossible – if these interests wish to do so and are united in that position.[18] We will consider problems of implementation below.

3.4. SOCIETY AS A FACTOR IN ENERGY

The energy available to a society is a major determinant of that society's behavior and potential; it is also true that society's values and decisions are crucial to how, and how much, energy is produced and distributed. For example, it is clear that to produce and make appropriate use of any but the most simple energy systems, the society must be able to provide the education

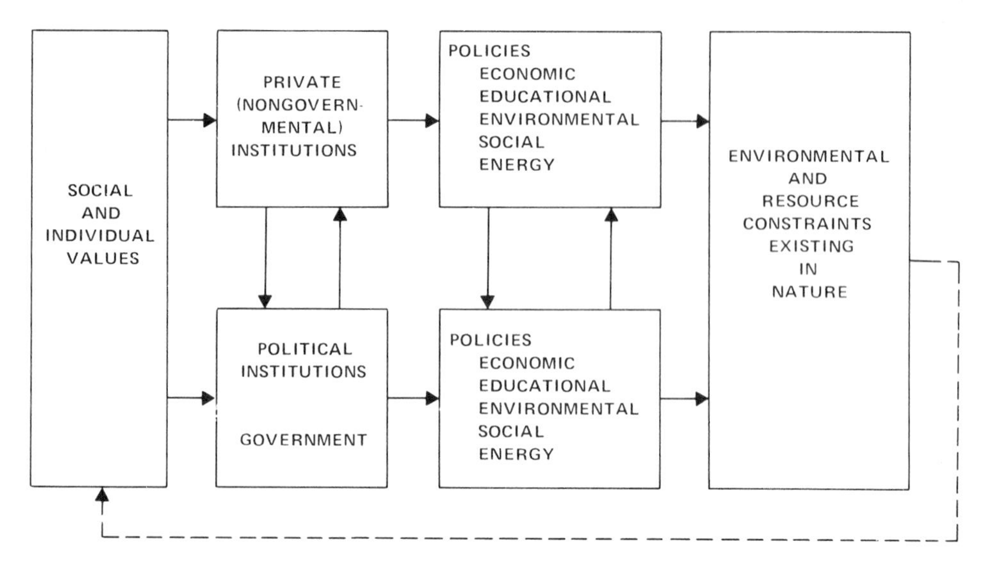

FIGURE 1. A model of the policy process.

necessary to operate them. Although the provision of properly trained specialists is fundamental, it is also necessary that the general public – the users – be trained in the proper and safe use of the energy forms supplied. Hospital emergency rooms can provide ample evidence that this education has its limits: even the American citizen, after many years of exposure to energy in several forms, still frequently plugs in his electric razor while standing in the bathtub or pours gasoline on a barbecue grill fire. Any new energy system must cope with the problems both of gaining public acceptance and of educating the public in safe use. Economic resources also must be provided by the society; this area is more extensively considered elsewhere in this volume.

The natural endowment of the society with geological and climatic or atmospheric benefits (minerals, sun, wind, water) is important. However, the society must be able and willing to extract and use these resources, and must judge the costs and benefits of doing so as compared to the costs and benefits of not doing so. Nuclear power is now caught upon the horns of this social dilemma: are the promised benefits of nuclear power worth the potential costs, economic and social, of nuclear power? The answer to such questions depends quite heavily upon the society's norms and values: what does the society believe to be good, just, and beautiful? what does it value more highly, safety or energy? Indian culture and the grandeur of the desert or the megawatts of the Four Corners power plants? To apply the simplified sociopolitical model of David Easton,[19] what will society demand in energy; to what extent, in what way will it provide the support to meet the demands it makes?

For hydrogen energy, one element of this question can be phrased in terms of how much added risk of injury from an unfamiliar energy form society will accept ("support") in return for an energy system which may be renewable and less polluting than petroleum. We have no satisfactory method of determining the answer to such questions except by practical experience, although some beginnings have been made on a methodology which could be useful here.[20] Instead, admitting our inability to derive politically and socially satisfying tradeoff solutions by technical or economic analysis, we have left the judgment of these difficult questions to our elected representatives. The several environmental protection laws, such as those controlling automobile emissions, are the political outcome of such socially perceived need; the technical or economic accuracy and adequacy of these laws can be disputed, but some such action was mandated by social demand. Hydrogen energy will face the same conditions: it is not enough to demonstrate the technology or prove the economics and expect the political system to respond favorably. Social norms will then be applied, and the outcome will be determined by political processes through action by interest groups, regulatory agencies, legislatures, courts – and perhaps the acts of foreign governments.

In the political process, some knowledge of American social norms will help us understand the probable patterns of action. For example, there is a clear preference, solidified in our institutions, for incremental rather than comprehensive problem solving.[21] This is functionally useful for an administrator, since decisions by increment reduce the amount of information required to make the decision, restrict the need for comprehensive action (which would perturb many other agencies and interests), and usually accord well with the atheoretic "practical" approach to problems which is generally agreed to characterize American society. But an incremental energy solution necessarily favors the status quo because it requires minimal disturbance of existing conditions. To the extent that hydrogen energy is to be a major departure from past methods, we can expect to find it delayed in implementation or completely obstructed. If there are ways to bring hydrogen into the energy system as an adjunct, a marginal supplement, or a partial substitute, to some degree the incremental preference can be utilized to smooth its passage. For example, it would no doubt be easier to gain permission to begin using hydrogen by proposing that small proportions be added to natural gas as a supplement to extend the life of that resource, rather than by proposing to lay entire new pipelines to and from new hydrogen plants and new consumers.

We may also usefully recognize the increasingly technocratic nature of decision making in many areas of society.[22] It is clear that complex energy systems such as hydrogen are not fully understood even by their initiates; the ordinary citizen is quite unable to comprehend the esoteric details. Our chief political decision makers are of necessity generalists, called upon to decide foreign policy one moment and utility tariffs the next. Decisions on energy policy are made against a background of general public opinion and in the framework of general social norms, but this leaves a great deal of flexibility. Because of this wide range of potential action and the general inability of most citizens (or political leaders) to fully understand the situation or the technologies, energy policy is in fact an elite decision area. Informed, interested "energy elites" will either make the decisions or exert powerful, frequently determinative, influence upon the political leaders who are nominally responsible.

We must also deal with at least one area of indirect economic impact: the political significance of the sheer size, complexity, lead time, and thus capital cost of energy systems, hydrogen or otherwise. The ordinary citizen, even the ordinary city council or small corporation, is not really free to make its own decisions on energy production and consumption. The need for access to massive captial resources naturally restricts the players in the game to a few major private corporations or to governments. Since very large-scale planning and coordination are required over lead times of a decade or longer, and since much money is involved, government will participate either as an original actor or because one or another private party demands governmental action to protect or advance its interests. There are many possible variations on government participation, present and future, from the "private power" companies regulated by government commissions to the state-owned Nebraska electric utility or Tennessee Valley Authority (TVA), with all manner of regulation, tax, and subsidy between. But it is certain that hydrogen energy will not become a reality without major government involvement to fund research, regulate and encourage or discourage implementation by taxes and subsidies. Politics will not permit governmental nonintervention in so critical an area; indeed, nongovernmental actors (such as businesses) will demand government intervention.

By participating in the energy system, government also imparts some degree of legitimacy to the arrangements used.[23] Legitimacy is to society what a triple-A credit rating is to individuals: it facilitates actions which could otherwise only be taken at much greater and perhaps prohibitive cost. There may be a great outcry if a private corporation charges a given price for its wares, but only minor grumbling if governmental actions sets a price as high or higher. An example is easily available: governmental tariffs on oil imports raise the price which citizens must pay for petroleum, but are generally accepted as "painful but necessary" in order to encourage more oil production in the U.S. If the American private oil companies were to raise their prices an equivalent amount in order to accomplish the same end, there would be outraged cries of profiteering and demands for *legitimate* government control of prices. Despite the lowered legitimacy rating of government since Watergate and Viet Nam, sufficient influence remains for governmental participation or nonparticipation to be of fundamental importance to hydrogen energy.

3.5 POLITICS AS A FACTOR IN ENERGY

The "correct" relationship between government and energy is subject to endless dispute, usually as a subcategory of the debate over the "correct" role of government in the economy. The parties to the dispute typically attempt to manipulate legitimating symbols such as "the public interest" or "private enterprise" to serve their interests at least as much as they depend upon the technical or economic merits of their case. The long and heated controversy over the federal government's proper role in generating electricity serves as a good illustration.[24]

There is no longer any point in arguing over whether or not government should be involved in energy policy; however, there may still be considerable value in debating the best way for government to be involved. The question is essentially one of defining roles and areas of competence.

Of the several major themes of American political life, the controversial relationship between government and private business is quite prominent. Today only the theoretical purists argue that while government has the duty to prevent monopoly, its role should end there. In the real world it is quite common for private business to pressure government to intervene in the economy with tariffs, price supports, tax writeoffs, or subsidies (e.g., historical subsidies to railroads, autos, and oil companies). It is now taken for granted that government will play a major role in any new enterprise which is high-risk and long-term, thus shielding private business from risk and using governmental financial powers and controls to allow long-term development. President Ford's September 1975 call for a massive governmental subsidy and guarantee program for energy development was unexciting partly because it is so ordinary.

Thus, there seems to be general agreement on broad definitions of role and competency: government is to provide certain financial incentives and guarantees, to use its control and planning powers to assure the creation or maintenance of certain economic and social conditions, and to provide appropriate regulation for safety and health; business is to perform the actual research, development, and operation under some form of contractual agreement. There is no apparent reason why this pattern may not be applied to hydrogen development.

It is worth noting that recent changes in both private and governmental organization may increase willingness and ability to work in the manner described above. Within the last decade the "energy company" has become a part of our industrial and political landscape, combining within one corporation the control of several varied energy forms. Davis[25] maintains that one difficulty in obtaining coherent energy policy has been that there are actually separate political arenas for each major energy form (coal, oil, gas, nuclear, electric) and that no overall policy is obtainable under these conditions. But this may be a historical condition rather than a future reality, given the growth of energy companies.[25]

Governmental structure has somewhat paralleled private evolution. For example, the Atomic Energy Commission (AEC) was once charged with the task of promoting and regulating only nuclear power. For various reasons, the AEC has now been fissioned into a Nuclear Regulatory Agency, which is no longer required to promote nuclear power but only to regulate it, and an Energy Research and Development Agency (ERDA), which is charged with research and development on all types of energy.

These two developments may be helpful to hydrogen energy's chances. With a broad spectrum of interests, the energy companies may see hydrogen as simply another energy form to be exploited, rather than as a potential competitor for the company's central interest in some single energy form. With its charter expanded to promote the development of energy resources in general, ERDA may be able to regard hydrogen as part of a comprehensive energy system rather than as merely one possible output of a primary interest (nuclear plants). But a caveat must be entered: because electricity and hydrogen have many similarities as secondary energy forms, the above-noted bureaucratic inertia may well favor electricity. Should this be so, it will be harder for hydrogen to break into this controlled arena than would be the case if energy policy remained largely fragmented among several interests and agencies.

However, within this broad range of agreement there is still room for much variation in detail. It may be that private enterprise and government will work in their usual division of labor, but it is also possible for government to directly enter energy production fields: government already participates actively in electricity, from municipal utilities to TVA. Government production may provide a yardstick by which to measure the productivity and prices of private concerns in the same field;[26] the same argument has recently been used to support the formation of a governmental corporation for oil exploration and production. Short of actual energy production, government may exercise virtual monopoly through a monopoly of part of a system or through heavy regulation, as in nuclear power. Given the added impetus of national security arguments for energy independence and the international financial problems of purchasing vast quantities of oil abroad, government will probably move steadily into greater production and regulation of all kinds of energy. With this background, designs for a hydrogen economy should assume a very large, possibly even exclusive governmental participation in development.

We need not assume that "government" refers to the federal government only. It is possible that states such as those of New England, which are poorly endowed with mineral fuels, may seek active participation in energy systems which would provide them with their own local version of energy independence. Some application of hydrogen might well suit their needs: for example, a wind-electric, ocean-thermal gradient, or nuclear primary source might use hydrogen as the energy storage medium or to replace waning natural gas supplied from elsewhere. Since implementation of any new energy system must begin somewhere, consideration should be given to localized and specialized applications based on comparative advantage, as well as to the more grandiose schemes for total systems.

3.5.1. Energy Independence?

Following the OPEC oil boycott and subsequent concern over high prices for imported fuels, some hydrogen energy advocates have spoken of hydrogen as an important component of some version of "Project Independence." It is difficult to see how hydrogen can make a very significant contribution to this goal, since it is a secondary form of energy which is necessarily derived from some primary source. "Independence" thus depends on finding suitable primary sources; hydrogen can at best be a useful component of some more basic system.

However, as a synthetic fuel, hydrogen may be useful at the margins. Several studies have cited areas in which hydrogen might replace petroleum fuels entirely or in part, such as commercial aviation.[27] Since many parts of the economy could be operated efficiently and economically on electricity (given a source of power to produce it), except for the unfortunate fact that bulk electricity cannot be stored like petroleum, an energy storage system based on interconversion of electricity and hydrogen might allow greater electrification of the economy and hence reduce petroleum demand. Of course, this would require that hydrogen be part of an energy system based on a nuclear, solar, coal, or some other primary source.

Although for foreign policy purposes hydrogen would seem a partial aid rather than a total solution, it may be that foreign policy will have a great impact upon the probability of implementing a hydrogen system. The symbols and reality of "national security" and "independence" are quite powerful, and are conceded by all to lie within the proper sphere of government. Certainly, if hydrogen advocates can design some method by which national independence in energy would be measurably increased, the demonstrated power of this argument would be sufficient to overcome the many barriers of economics, interest rivalries, and public doubts. It is not yet clear whether such a design will be forthcoming.

3.5.2. The Role of the Public

We come, then, to a consideration of the role of the public in the politics of energy, especially hydrogen energy. In a country which prides itself on being democratic and which maintains institutional structures which encourage at least some sensitivity to "public opinion," no major policy can be formulated and implemented without at least symbolic consultation with the public. Further, since energy is a fundamental component of the life of every citizen, no policy which the public cannot implement can succeed. For example, it would be futile to propose banning private automobiles except in the most dire straits conceivable, simply because many citizens depend on automobiles not only to commute to work but to commute long distances to the grocery store. The rural Kansas citizen cannot reasonably be serviced by mass transit; he must drive a private car or revert to the horse. However, certain restrictions on use, either directly or through rationing, can be implemented. The question of practicality then becomes one of political feasibility: can the decision makers involved recognize the problem, formulate a policy, and remain in office long enough to implement the policy, secure public acceptance, and make it work?

Specifics of public participation will be discussed below in the context of hydrogen energy. The general position of the public is rather easily stated, however: the public expects to have nearly all the energy it "needs" in a convenient form at a tolerable price, because it has become accustomed to that condition. The American public now disposes of vast per capita energy resources; the history of energy in America is of increasing availability of resources through "normal" economic workings or specific public policy choices such as the TVA or the Rural Electrification Administration. The public shows little preference one way or another on such questions as public vs. private power, although political and economic elites may become quite agitated and may seek to arouse the public by appeals to popular symbols such as "government interference." Essentially, the public expects to have adequate energy; if it does not, it is more likely to complain to government and to demand compensatory action than to examine its priorities and values. Davis has noted that there is a long-standing trend, a "flight from the market" in which both energy companies and consumers have cooperated in shifting from the price system to one of governmental guidance. This produces what he calls "the primacy of politics . . . [a] great unexamined presumption."[28]

Environmental politics show a characteristic development pattern which apparently applies in energy politics also. The pattern begins as established policies and practices, even social norms, are increasingly questioned by small but growing groups of opponents. The opponents typically organize around some specific local issue, but umbrella organizations grow rapidly to tie the fragments together or to provide active and perhaps "expert" assistance. Media attention expands the issue, and if no satsifactory local settlement is reached, conflict moves on to state and/or national levels. If the same general category of issues develops in several areas and gains public recognition, it begins to obtrude upon politics sufficiently to produce legislative proposals, administrative regulations, and court cases. The issue may then fade, or it may itself become part of the establishment in the form of national policies, as have the various environmental laws, vehicle safety and emissions controls, nuclear power agency reorganization and controls, etc. This pattern seems to be international and to affect even nondemocratic states.[29] We should note that action is characteristically generated by elite groups, but against a broad and shifting background of public interest.

If we may assume that this pattern will hold in energy policy, our attention should focus on the public less as an initial actor than as a recipient and follower of policy, with the important caveat that changing any of the "traditional" boundaries of public activity would require more time and energy than will any policy which remains inside those boundaries. Thus, we should look to elite groups for initial actions and reactions. They will set the agendas for discussion, develop the broad alternatives, bring pressure, and defend their proposals.[30] In a policy area as highly technical and complex as energy and energy policy, claims to technical expertise and a significant information base will be critical resources, but ultimate implementation requires public acceptance. This suggests that policies which follow a line leading from laboratory research and development through expanding scales of demonstration will be necessary. Of course, such a long-term program also provides any opponents with the maximum time and opportunity to torpedo the program outright or to render it innocuous. There is a perhaps apocryphal story of a state legislator who could not convince his colleagues to abolish the state's most severe sentence, death by hanging. He was, however, able to insert a rider in the budget bill which forbade the state to purchase rope.

3.6 HYDROGEN AS A SPECIAL CASE OF ENERGY, POLITICS, AND SOCIETY

The technology and chemistry of hydrogen energy are the concern of other authors in this series. Clearly, if these technical matters cannot be properly handled, the social and political impact of hydrogen will be largely a study of energy system maintenance as yet another contender fails in its challenge. It is necessary here to make certain assumptions:

1. That technical problems of storing and transporting hydrogen are soluble with technologically acceptable processes and methods, and with at least marginally acceptable economic costs;
2. That no presently unforeseen synergistic problems will arise which will make the pieces of hydrogen technology individually acceptable but will make a "hydrogen energy system" an overall failure;
3. That the technically desired methods of production, transport, and storage do not run afoul of irremediable environmental objections.

Although the point has already been made, it is worth reminding ourselves that while hydrogen may be a marvelous fuel with exceptional environmental qualities, it is not a primary energy source. If we find that hydrogen can be produced economically only by coal gasification, for example, hydrogen becomes prey to all the varied objections which may be levelled against anything having to do with coal mining and use. In effect, no decision to implement hydrogen use can be made without having previously decided upon the economic, environmental, safety, legal, political, and social acceptability of its generator. Having noted this point, we can proceed to discuss hydrogen as we might electricity, another derivative energy form which can be made from a variety of primary sources.

Environmental arguments are probably the strongest in the current armory of the hydrogen advocate and need no extensive recapitulation for readers of a volume such as this. If we can solve the problems of energy to generate it, hydrogen itself has the virtues of clean combustion and almost infinite replenishability rather than the familiar decay curves for supplies of petroleum, uranium, and coal. However, we should note that these virtues apply with equal force to electricity. The problem, then, is the standard one of comparison: how does hydrogen differ from its major rival, and what tradeoffs may be or must be made between them?

Hydrogen would seem to have considerable advantages over electricity in both storage and long-distance transmission. The advantage is both technical and environmental. Electricity cannot now be stored in bulk, and such substitutes as pumped-storage hydroelectric systems are open to serious environmental objections. Transmission of electricity over long distances is now less energy efficient than gas transmission by pipeline, and is also environmentally less desirable. For equivalent energy volumes, current overhead electric transmission lines require rights of way as much as an order of magnitude larger than for a pipeline (150 to 200 ft wide as compared to 15 to 20 feet); further, the pipeline is out of sight under the ground and causes minimal restrictions on surface activities, while electric transmission lines are seldom an esthetic delight and may restrict nearby construction. Since the right-of-way area for transmission lines is already in excess of 10,000 mi^2 (larger than Connecticut, and growing) – with increasing opposition and rapidly mounting expense to both utilities and consumer – the land-use impact of overhead electric lines will further inhibit their extension.[31] It is not yet clear whether the same statements can be made about local electric distribution compared to gas lines, since the technology for short-run underground transmission has improved somewhat, but it is clear that overhead high-voltage transmission lines, both rural and urban, will meet severe obstacles in the form of cost, environmental protests, and land-use controls in general.[32]

Long-distance high-voltage lines have already been the subject of political and court battles in several states. Because these struggles always consume several months or even years, the cost of the line can be raised considerably as capital lies idle, inflation continues, markets are unserved, and legal costs increase. Further, even if building is eventually allowed, it may be along a revised corridor and with added esthetic or safety requirements which further increase costs. For as long as storage and transmission tradeoffs remain as they now are, hydrogen should be politically and perhaps economically preferable to electricity for long-distance transmission, and preferred over pumped storage for peak-load facilities.

However, hydrogen has a political disadvantage which is well known to both proponents and opponents: the safety issue symbolized by the "Hindenburg Syndrome." This presents a problem which technical excellence can help resolve, but which is really beyond technology. It is precisely the danger potential which is at the same time the value of any energy source: gasoline can burn and blow up, which is why it is both so useful and so dangerous if wrongly used. No energy source can ever be made completely safe; nor has society demanded of its energy systems that complete safety be assured. The difficulty is that any new energy form must overcome barriers of unfamiliarity, which may both make accidents more likely and increase the shock value of any accident.[33]

We do not know what level of danger society will accept in return for the benefits of an energy form. This question is the root of many current delays, cancellations, and costs in the nuclear power programs of many utilities. Opponents need not prove that massive reactor failures will occur, nor that such a failure will kill 100,000 people, in order to arouse public opposition. It is not enough for proponents to show that the probability of a massive failure is "only" 1 in 10,000 per plant per year. Society, including most political leaders, is sufficiently unfamiliar with the technology and sufficiently afraid of real or imagined dangers to react with great concern to even the smallest doubts about reactor safety.

Nuclear energy is in effect in an education and familiarization process as society learns about a new energy form, thinks seriously about whether there are better methods of producing energy, and comes to some conclusion on policy. Education can be expensive and time-consuming, but is is supposed to result in a higher-quality life. If the public and its political leaders become more familiar with the idea of nuclear power, and if the arguments and safety measures seem adequate, we will develop major programs of nuclear power; if not, we will not.[34] The decision process in politics is ultimately subjective, although objective data play a part.

Hydrogen energy will also face the obstacles of unfamiliarity and a negative image. It may be taken as a given that expense for "unnecessary" safety measures must be incurred, that troublesome regulations will be imposed, and that timetables will be optimistic. Some strategems will be suggested below which may help to minimize these conditions, but there is no eliminating them.

3.7. CREATING A DECISION FOR HYDROGEN

For purposes of discussion, we will assume that technical requirements for the economical and safe use of hydrogen (as determined by public willingness to accept the risks) can be met. When this point is reached, some way must be found to present hydrogen as a viable alternative to people who have the authority and ability to make policy.

3.7.1. The Context

As noted above, anything beyond very narrow, specialized uses of hydrogen will necessarily involve massive sums of money and the interests of several established groups: governmental regulatory agencies, public and private utilities, energy companies, unions, states with either energy resources to sell or energy needs to be met, etc. The technologies and ideas involved will be new for most political purposes, although they may be old to practitioners of hydrogen research. Only the fairly narrow stratum of the public interested in such matters, those whom V.O. Key called the "attentive public," will pay much attention.[35] There is a larger public, of course, but if it responds at all, the response will be based more on such keys words as "energy," "price," or "jobs" than on hydrogen's peculiar virtues. Thus, the nature of the technologies, the general problem of energy, and the financial facts involved will make pro or con decisions on hydrogen a matter for elite groups rather than for mass participation. A decision to implement hydrogen as a utility energy storage system, for example, will no more be subject to public "election" as a decision method than would a decision to switch from natural gas boiler fuel to coal. Decision-making groups must bear in mind the possibility of activating latent public opinion, and counter-groups or portions of the energy elite with conflicting interests may certainly oppose the decision; however, the question remains within the attention and control of only a small minority of the total population, unless one or more of the parties seeks to improve its own political position by "going public," seeking to gain leverage it does not have in intra-group bargaining. California referenda on energy and environment issues are examples of the latter strategy.

We may, then, reasonably expect that the elite groups concerned with the original decision will be people who have heard of the Hindenburg many times (they may even belong to the Hindenburg Society) but to whom that particular image is a datum rather than an emotional experience. That datum may well cause them to be particularly sensitive to safety questions and to insist upon rigorous safety measures – thus, expenses will increase – but it is not likely to engender emotional opposition.

Hydrogen then will be accepted or rejected initially by elite groups, likely to be more affected by pragmatic questions of feasibility, economics, and institutional interests ("bureaucratic politics") than by either technological razzle-dazzle or unfounded fears for safety. They are also likely to have at least a basic grasp of technology, some definite ideas about economics, and some career and corporate momentum to preserve. As members of an elite segment of the American population, they may be somewhat more aware of the limitations of both technology and organization than is the average person, but they will not be immune to normal U.S. cultural values: the acceptance of technological solutions, a willingness to adapt to new conditions and to experiment, and a denial of the proposition that some problems have no good answers.[36]

The decision must be made within certain constraining parameters, or the parameters must be changed. Either is possible, but the inertia of institutions, policies, and habits is such that remaining generally within established boundaries is preferable in decision making. Halperin observes of any bureaucratic system that "it is basically inert; it moves only when pushed hard and persistently. The majority of bureaucrats prefer to maintain the status quo, and only a small group is, at any one time, advocating change."[37] Thus, some of the cultural values – innovation, change, problem solving – clash with some of the standard operating procedures of political and social institutions. The problem of decision and implementation for hydrogen energy is, then, very similar to problems faced by any "new" idea, complicated or eased by whatever factors may be developed by the special characteristics of the energy system: technological, economic, environmental, etc.

"Selling" the system to the public will be required at some point, unless the system is so inherently restricted to an operating elite (e.g., a utility using off-peak power to generate hydrogen as an energy storage mechanism, storing the hydrogen quietly on its own property) that the public is probably not aware of its existence. Any of the large-scale applications – commercial aviation, automobiles, or home appliances – which necessarily involve the general public must both convince the public that the system is useful and safe and teach the public to operate it properly. Further, it will be necessary to obtain public support for the revision of building codes, financial requirements of conversion, local government cooperation, and many other requirements.

We know much less than we should about social response to technological innovations. The growing field of technology assessment, the inclusion of SIA as part of the required environmental impact statement for new projects, and the burgeoning use of "science advisory boards" at even the state and municipal level indicate that this lack is recognized and that action is being taken to remedy it.[38] We do know quite a lot about how to "sell" a new system, thanks to the long labors of Madison Avenue's legions, but how to cope with the problems arising after the sale – teaching safe use, adjusting differential social impacts among socioeconomic groups, etc. – is unfortunately still an undeveloped area.

There are two main conventional methods for dealing with these problems: education, both formal and informal ("propaganda"), and administrative or legislative regulation and compensation. Formal education already includes both explicit and implicit instruction in the safe use of dangerous materials and energies. Informal education, of the kind utilities have performed by including with the monthly bill a reminder that one should not check for natural gas leaks with a lighted match, could certainly be used for hydrogen as for other energy forms. Because hydrogen would be a new and unfamiliar source of danger, and because prevention of public injury is of interest to state governments, it should not be difficult to prepare and introduce into school curricula in affected areas a "short course" on hydrogen safety.

Regulation and compensation are more difficult. Compensation may be needed if a proposed hydrogen system will in some way adversely affect some social or economic group, just as compensation in several forms is now being made available to persons adversely affected by coal-mining operations. Because state and particularly federal agencies have become very active in this arena only in recent years, the older energy forms have been dealt with on an incremental basis and were able to fight back from the position of strong, vested interests. Hydrogen would face the possiblity of *de novo* regulation and compensation requirements, applied with full rigor from the beginning. This might well raise the economic costs of the system, but should also reduce the probability of significant accidents – a point which could prove decisive, since an unknown fuel with the flaming Hindenburg already in its past could be severely penalized or even stopped in its tracks by a major disaster early in its development. We may suggest an analogy to nuclear power: although nuclear power plants have been remarkably free of accidents causing injury or large-scale public danger, safety is still one of the major arguments used by opponents and is powerful enough to place economic penalties and real restraints on nuclear facility construction and use. Had there actually been a major nuclear failure, or if one should occur in the near future, we would expect potentially crippling consequences as regulation and compensation costs and requirements leap to new levels. To avoid similar penalties, it may be well to develop a thoroughgoing safety and educational program in advance of hydrogen implementation, secure governmental approval and incorporation of the plan, and thus begin operation with a known quantity rather gambling on after implementation developments. Such preparation would also serve as evidence that safety is being taken seriously and thus would reduce the ability of opposition groups to fasten on this issue.

The critical question here is essentially quantitative: how much public acceptance is needed in order to make a hydrogen system acceptable? The unfortunate fact is that we cannot give a useful quantitative answer; about all we can do is to say that the level is perhaps not very high because of the essentially elitist nature of the decision process and the probability that the first uses will be of a type which involves a fairly small number of people accustomed to highly technical operations. When a genuinely public application is proposed for large-scale use, we will need to seek specific responses to the proposal, but will have a much better data base due to prior experience, and should be able to ask more productive questions.

3.7.2. The Process

Students of public policy have attempted to spell out several stages of policy making and operation. Policy can be defined either as what is operational, or as what is *declared* to be the goal or method. A good illustration is found in recent federal energy policy, in which nearly all major figures and organizations have *declared* that American policy should be unified and should be directed toward lessening dependence on external energy sources and increasing our supplies. The operational policy has remained anything but unified, has increased our supplies little, if any, and has actually allowed our dependence on external sources to increase. It is well to remember that declarative policy and operational policy may be very different.

Even so, the first step in policy must be the creation of a decision. The decision may be specific or general, may be taken at virtually any level, and may be explicit or implicit, but it must be made in order to begin movement and build pressures which lead to later actions and declarations. To create that decision, and more specifically to create the "right" decision by the "right" person or organization, is the business of political lobbyists, legislators, and executives. Since any significant use of hydrogen will require action – or benevolent nonaction – by public officials, we must consider how to create a decision and then how it might be implemented as policy.

Public policy is made, according to the most commonly accepted interpretation, by the clash of interested parties and a resulting synthesis which is translated into legislation or administrative regulation. The initial synthesis immediately becomes the base line from which the contending groups seek to move the policy further toward their preferences. It is rare indeed that an issue is "settled" in anything more than a temporary, partial sense.

Policy may be made in either or both state and federal administrative agencies, legislatures, or courts. It is not unusual for different participants to make different, even conflicting policy, such as federal agricultural programs which assist tobacco growers while federal health policies force cigarette makers to warn that "smoking may be dangerous to your health." Thus, it is frequently difficult to say that there is a general policy, and impossible to say just what that policy is. Further, policy is in constant flux as interests wax and wane in power or access to policy makers and as individuals and agencies change in government. That this is a fact, and recognized as such, is shown quite adequately by the large donations to political campaigns which are common throughout the business world and by the extensive lobbying efforts supported by both commercial and noncommercial groups.

Much of the public support for a hydrogen

economy will come from two basic sources: well-founded concern that fossil fuels are growing increasingly expensive and scarce, and a desire to minimize the environmental insults created by energy systems. Neither of these is likely to lessen soon, although emphasis will shift periodically. General public interest in these problems may be expected to follow what Downs has called "the issue-attention cycle."[39] The cycle beings with a "preproblem state" in which a problem exists but is known only to the experts, then proceeds to a second stage of discovery and alarm in which the public assumes that a) there is of course a solution to the problem, and b) somebody ought to do something to find the solution. At this point, government becomes a focus of pressure from the public. The recent sudden public discovery of an "energy crisis" which was at least 20 years in the making is a good example of the first and second stages. The third stage is a "morning after" phase in which the public and the government begin to realize that the solution, if any, will cost them something – money, injury to other values, inconvenience, etc. – and to look for some acceptable tradeoffs. At this stage, the faith in technology implicit in the second stage becomes operationally important, because the public may readily pursue some new process or device which promises somehow to reduce the costs. It is perhaps appropriate to note that the hydrogen economy argument has more than a little of this hopeful optimism and to suggest that the enthusiasm for hydrogen may decline in the fourth stage, which is one of growing realization that "there ain't no such thing as a free lunch." In this stage, there is a tendency for the disillusioned public to "turn off" and become progressively less interested. Lastly there is a "postproblem stage" in which general public disinterest exists but may not be very significant because of what happened in the second and third stages: creation, on the wave of public concern, of institutionalized programs directed toward "solving" the problem. Once institutions are established, staffed, and built into the budget and organization chart, they become a continuing vested interest, actors in the policy process with a built-in reason for keeping alive the problem which brought them into existence and with a mission to solve the problem. For example, we may cite the Environmental Protection Agency (EPA) as a product of just such a cycle.

3.7.3. The Requirement of Representation

The "hydrogen economy" may well be an idea whose time has passed. To date, the institutional response to the energy crisis has been almost exclusively in terms of increased emphasis on the conventional, known sources and methods. ERDA's increased funding of exotic methods and such evidence of interest as the rather wide-ranging hearings before the subcommittee on Energy of the Committee on Science and Astronautics in the U.S. House of Representatives are minor compared to the increased efforts in already institutionalized programs such as nuclear power. It is of considerable significance for the hydrogen economy's future that ERDA, the designated manager for federally-sponsored energy research and development, has at present a low interest in hydrogen.[40]

We should not be surprised that the attractive systemic arguments for a hydrogen economy have not been seized upon as policy. Huntington[41] has observed that major policies "are not the product of expert planners rationally determining the actions necessary to achieve desired goals. Rather, they are the product of controversy, negotiation and bargaining among different groups with different interests and perspectives." Since the hydrogen economy does not have a significant institutional representative, and since even those groups which might find hydrogen useful (e.g., petrochemicals, energy companies, utilities, etc.) may consider it as merely one of several possible alternatives, a willingness is likely to exist to bargain away hydrogen alternatives in order to obtain some better-represented system.

Until such time as some major interest group decides upon hydrogen as its preferred alternative and begins to press vigorously for its choice, there is not likely to be more than "back burner" research and development support by either government or private companies. If there is only limited support, other more established alternatives (coal gasification, electricity) will continue to lead and perhaps even draw further ahead in the race. In June of 1975, the Subcommittee on Energy Research, Development, and Demonstration of the House Committee on Science and Technology held hearings on hydrogen energy; none of the organization-sponsored witnesses represented organizations with a primary interest in hydrogen. Such evidence suggests that although hydrogen will very likely be more widely used for specialized applications or in some industrial areas, it is not politically realistic to expect a full-blown "hydrogen economy" for many years. To advance the data would require not merely further technological development or rising fossil fuel prices, as is usually suggested, but the creation of a powerful and politically active interest group dedicated to the vigorous advancement of hydrogen as "the only way to go." No such organization is now visible.

Any organization which determined to push for a hydrogen economy would have to find supporters in order to build an alliance leading to policy success. Who might these supporters be, and why?

Environmentalist agencies such as EPA and private groups such as the Sierra Club or Environmental Defense Fund would seem to be natural supporters for hydrogen because of its potential contribution to clean air and a recycling economy. Unfortunately, it is too early to be sure of this. The environmental Achilles heel of hydrogen is the source which produces hydrogen, which is perhaps made more serious by the fact that the unavoidable conversion losses will require greater input from the source than if a more direct method were used. For example, Plass concludes that any depletable resource used as the energy source in a hydrogen economy would be depleted more rapidly than in a nonhydrogen economy based on the same resource. Even more signficantly for the environmentalist, in at least some systems the environmental damage over the total system may be higher for hydrogen than for the alternate.[42] It is too early, and the technology and requirements are too little known, to be able to say whether Plass's conclusions are accurate, but they do suggest that the presumed "clean energy" attractions for the environmentalist must be taken more soberly and cautiously. Thus, the environmental movement may aid – or it may oppose – hydrogen energy systems, depending on whether the overall systems proposed are part of the solution or part of the problem of environmental cleanliness. We must be sensitive to system net energy efficiency and to system net environmental impact.[43]

Public and private energy interests (energy companies, utilities, TVA, etc.) will no doubt choose to support or ignore hydrogen on a largely economic basis. However, economic calculations increasingly must include internalized social costs, especially health and environmental protection costs, which tend to be determined by public policy decisions rather than by the market or corporate policy. Thus, economic burdens of various energy systems will be heavily affected by public policy; indeed, choices may be virtually forced by combinations of taxation, subsidy, and regulation costs. One can easily construct any number of scenarios in which governmental decision would compel a particular choice, but one cannot now predict which scenario is the more probable. However, one can say that governmental choice is strongly influenced by the arguments and bargaining power of private groups (including "independent" government agencies, such as TVA).[44] Support or opposition by powerful energy interests probably depends on whether environmental and social cost studies can present a compelling economic rationale for choosing hydrogen over some other energy solution. The approach begun by Plass could be further refined and elaborated to test whether such a rationale is feasible.

Provided that the primary source of hydrogen is from a domestically controlled resource, a hydrogen economy would allow a high degree of energy independence for the U.S. or any other country, a point which Japan and Western Europe have also noted. The appeal to utilize U.S. coal resources to fuel an electric-syngas system is based on essentially this level of argument; it has gained some credence but is offset by the known environmental costs of coal. A dual system of electricity and hydrogen, both produced from domestically available, plentiful, nondepletable resources, may be environmentally preferable and could bring many countries close to the national energy independence which three U.S. Presidents have announced as a high-priority goal of U.S. public policy. For both international economic reasons and the broader goal of a nationally secure energy system, hydrogen is not the only answer, but it is potentially one of the better ones, and as such should appeal to a powerful governmental and public emotion. The task of hydrogen economy advocates will be to convey this message to the public in such a way that support is secured.

3.8. PILOT PROJECTS

It is routine in technological development to build and operate model facilities before scaling up to full-size systems. A number of these pilot projects of varying scale have been suggested for the hydrogen economy, covering industrial, residential, transportation, and other uses. Most of these proposals are intended to prove the technology, although some are concerned with proving social and political as well as economic acceptability.

From a political strategy perspective, any pilot project should seek to accomplish two ends; first, the creation of programmatic vested interests and inertia which will carry the project forward in the face of opposition or apathy; second, the disarming of competing vested interests and the dissipation of inertia in competing programs.

Within the past 3 years, a combination of several factors has led to a rapid rise in the attractiveness of solar power. Building on this, it has proved possible to obtain greatly increased solar research funding (Public Law 93-473). Probably of greater long-term political importance, however, is that legislation has provided for a Solar Energy Research Institute, for which several states are eager to provide a home. John Sawhill, then Administrator of the Federal Energy Office, had commented that solar and geothermal energy "tend to get left out because they don't have strong pressure groups behind them."[45] This institute will become a group of "inside men" with a vested interest in maintaining and expanding solar energy systems; the projects funded by this and other institutes and agencies, as well as the research divisions of private businesses, will all add to support for solar work and will increase access to the policy process. The states and communities in which the agency and its projects locate will lobby for continuing and expanding solar work. To the degree that they are able to secure attention for solar power, they will compete with alternate energy systems; this competition produces patterns of both mutual conflict and mutual support. Solar energy, now with a political base, is more likely than before to become a significant portion of the national energy system. To be successful, hydrogen must follow a similar route; in fact, pilot hydrogen projects could seek to show how they can fit into ongoing solar projects while also working to establish some independent political position, in case solar energy should prove somehow limited in usefulness for a hydrogen economy.

However, the very act of seeking to establish research and development programs and pilot projects inevitably means that hydrogen will enter the active political arena, if for no other reason than that the research and development budgets of both public and private institutes are limited and will be fought over by competitors also seeking funds. In such combat it is seldom enough to have the "best" program as one enters the legislative struggle; if it were, politics would virtually cease to exist and sheer technical excellence would dominate all such competition. Clearly, that is not about to happen.

As part of the political combat, hydrogen advocates will find themselves forced to contend with multiple sets of requirements and prohibitions placed upon any pilot projects. Many will no doubt be due to "good" reasons of technical monitoring or experiment, concern for public safety and environmental protection, etc., and similar kinds of restrictions and compulsions will be placed on all new energy programs. However, in the cockpit of politics it is not unknown for program opponents to manipulate these and other factors in such a way as to impose an unsupportable burden upon the program being explored and thus to kill it by indirection. It is surely true, for example, that some persons who accept the "no growth" argument would seek to require a very high level of environmental protection regulations for any new energy system, attempting to obtain either a maximal goal of stopping the program or a minimal goal of at least minimizing the inevitable damage if the program is implemented. For hydrogen, as the existence of the Hindenburg Society explicitly recognizes, safety will be a vulnerable point upon which both neutral parties honestly concerned with public protection and the opponents of hydrogen, for whatever their reasons, may combine their attacks.

For anyone seriously concerned with advancing the cause of hydrogen, covering one's political flank must take high priority. In politics as in the martial arts it is sometimes possible to use the opponent's attack against him. If a proposed pilot project for hydrogen can include a detailed, comprehensive, "oversafetied" system of protection for operators and the public, and if environmental protection can be made a major feature rather than an appendix of the proposal, safety and environmental protection may be converted from liabilities into assets.

The essence of this point then is simply that one should do one's homework and be prepared with strength in what would otherwise by a weak spot, such as the safety of hydrogen. It will do little good to protest "unfair" or "unrealistic" safety requirements, since safety is so thoroughly a judgmental matter. Further, to refer again to nuclear power and the massive safety requirements placed upon power reactors, it is worth noting that several serious "accidents" have happened even with the supposedly redundant, safety-engineered systems now in use. We can, of course, only speculate, but is may be useful to consider the potential political impact on the future of nuclear power if the "incident" in the Fermi #1 breeder reactor near Detroit had led to a major accident in which several hundred or thousand people were killed and injured.[46] At minimum, there would surely have been drastic increases in insurance fees charged, or private insurance might have become unavailable for nuclear power, and investment capital might well have dried up, while federal and state regulations would surely have become far more stringent than they are. Repeal of or refusal to extend the Federal Price-Anderson Act, which provides nuclear insurance and sets limitations on utility liability for nuclear accidents, might have effectively strangled the industry. In effect, we may speculate that the safety regulations which have created high costs and much carping about "over regulation" in the nuclear industry are a major reason that nuclear power today still has some future. Hydrogen advocates may not exactly welcome stringent, complex, high-cost safety requirements, but it would be appropriate to consider them as one of the necessary components of a hydrogen economy and to go several extra miles in designing safety measures. Government will surely insist upon high safety standards; there will be no ducking the issue, because competing energy interests will not permit it.

The "where" of pilot projects should also be considered, along with the "what." Pilot projects should be located in or have maximal effect in areas where competing power systems are weakest. Pipelined hydrogen in several New England states, for example, would potentially replace the natural gas which is already in short supply in that region, and would possibly ease many of the political tensions which are already evident between states which are natural gas suppliers and those which are consumers. As the TVA proves, it is possible to build regionally-oriented programs which then develop considerable political muscle as the states with interests in the program make use of the leverage inherent in a federal political system. One study has already proposed a pilot project in such a location as Hawaii or some other fairly remote island, where a hydrogen system substituting for petrochemicals may well have favorable economic results when compared with fossil fuels.[47] If it is possible to couple hydrogen to such island resources as ocean-temperature differentials for generation of basic power, preserving the important "clean environment" factor in a tourist-affected economy, a suitable pilot project on the island state might be very attractive, and would again take advantage of national access through the political channels of federalism.

3.8.1. Arranging for Regulation

One factor which may be of crucial importance is regulatory agencies which will control hydrogen pilot projects or later permanent programs. A single agency, such as the Federal Power Commission, might acquire regulatory powers if the project is a limited one similar to natural gas pipeline use; a large-scale project involving, for example, a nuclear power reactor which electrolyzes water and pipes hydrogen across a state line to be used by utilities for peak-load generating capacity might find itself regulated by a combination of several different agencies, both state and federal. As with many other hydrogen questions, there is a limit to useful comment on this area until a specific system and its alternates are proposed. We do know that some federal regulatory agencies are closely intertwined with the interests they nominally regulate and could be expected to favor whatever those interests favor. We also know that a system subject to regulation by several different agencies has both increased complexity and, under some conditions, increased opportunity to maneuver from one agency to another, while invoking the support of one agency against the other.[48] Part of the political planning of pilot projects or large-system development should, therefore, include consideration of how best to design the system within advantageous agency lines of jurisdiction or interagency "gray areas" and to minimize probability that a hostile agency will be made part of the regulatory team. Again, such planning must await clarification of production, transportation and use preferences so that specific information can be brought to bear on specific proposals.

3.9. SOME INTERNATIONAL CONCERNS

A hydrogen economy of any size will have definite, perhaps very significant, consequences for the international position of the U.S. To the degree that hydrogen might allow greater energy autonomy, not only the U.S. but any other nation with hydrogen economy capabilities would become less dependent upon fuel-supplying nations. Because the U.S. and its major allies, both the NATO nations and Japan, are the world's largest consumers of imported fossil fuel, it is probable that the development of a hydrogen economy would represent a very significant gain in the international strategic position of the U.S., Western Europe, and Japan. The desire for energy autonomy is in part responsible for the current petroleum-deficient state of the U.S., in that past "national security" policies encouraged the production and consumption of domestic oil supplies at an accelerated rate, thus "draining America first." The national security argument is clearly powerful, although its consequences are not always fully understood. It would be appropriate for a full-scale study of the national security consequences of a hydrogen economy to be performed at an early date, as soon as we can develop some reasonably useful estimates of the scale and sector impact of a feasible hydrogen system.

It should be noted that there is no reason to believe that the U.S. would have a monopoly on the possibility of a hydrogen economy. It is possible that an intensive research and development effort would give the U.S. at least a temporary corner on the technology involved, which could no doubt be used to both the political and financial benefit of the U.S. But we should consider what other nations would be affected, and how. For example, the OPEC nations would clearly be affected, and their response strategies are not clear. Because of the potential for reduction in dependence on external sources of fuels, such fuel-short but technologically advanced nations as Germany, Italy, or Japan may adopt the hydrogen economy solution and become centers for both technological development and pilot projects. Some of the developing nations, increasingly hard-pressed to find the money to pay the growing cost of oil, petrochemicals, and artificial fertilizers, might find hydrogen a boon in dealing with their needs. If so, there would be additional shifts of international positions, financial flows, and trade. Thus, the magnitude and direction of the international effects of a developed hydrogen capability is not all clear; it should be studied.

As one brief example, consider Japan. If it is possible to develop an eclectic economy of both hydrogen and electricity, generated from such domestically available resources as geothermal areas, ocean-temperature differentials, or solar installations, Japan could significantly reduce its current extreme vulnerability to petroleum boycotts and prices. If it chose to do so, there would be a very significant impact upon petroleum prices, shipping, international finance and trade, and perhaps even military alliances. It can be maintained that if the "limits of growth" models are even marginally correct, Japan has little or no future in a petroleum-fueled world because of the economic, political, and military restrictions imposed upon it by geography and technology. Thus, an economy entirely or in part based on domestically-controlled resources and energy forms (hydrogen?) would be a key to maintaining Japan as a significant actor in international politics and finance.

It is, then, possible that a hydrogen technology is part of the key to a "Project Independence" not only for the U.S. but for other nations. It is not clear whether this is unequivocally a good thing, for either the U.S. or other nations, in part because it is not clear what the consequences of such an unlikely (at least for any near-to-middle-range future) attainment would be.[49] But this takes us from hydrogen into a far broader range of questions.

3.10. CONCLUSIONS

Someone has said that "we should all be very concerned about the future, because that's where we'll spend the rest of our lives." Unfortunately, the ability of social scientists to predict the future is certainly no more, and perhaps less, reliable that that of technologists. But society and politics are highly interactive, and technology is a part of both cause and effect in social and political events. A useful example is the "pocket slide rule" calculator, which recent developments in electronic circuitry brought to reality virtually overnight, while society lagged behind. Educational institutions still provide little training to allow their graduates to exploit the full potential of such calculators, but do find themselves struggling with practical problems in high school and college physics classes where it must be decided whether calculators can be permitted or "old-fashioned" hand calculations and slide rules must be used. There is even something of a moral problem, in that wealthier students who can afford a "pocket slide rule" calculator have an edge over poor students who cannot. Should there be a "welfare calculator" program? Neither the economic impact, educational importance, nor such ethical and practical questions as the differential social impact of pocket computers were predicted with any clarity as little as 10 years ago.

"The hydrogen economy" is, at this moment, a very amorphous set of ideas and vigorously disputed probabilities and designs, and is, therefore, hardly a suitable subject for rigorous analysis by the tools of SIA, particularly since it must be admitted that those tools are not yet sharp. Both the technology and the social-political-economic impact of hydrogen would be measureably advanced by agreement on one or two model systems, preferably operable as a small-scale but socially significant pilot model, so that a number of questions could be dealt with in a context of parameters allowing useful measurement. Until that time, any "social impact" studies of hydrogen must in honesty admit to a potentially very large and quite nonstandard error of measurement. Thus, this particular contribution tries for little more than a cataloguing of points which designers of a model system should seek to incorporate as far as possible.

This author's suggestion would be for a system which somehow involves solar power. This strategy would tap a rapidly developing technology with growing public and private support, and offers at least the possibility of enlisting the vigorous activity of environmental advocates and the continuing general public attitude of support for environmental protection. A range of solar power options are suitable for hydrogen production, but a large-scale and glamorous solar system would maximize hydrogen's visibility as a valuable energy carrier.

Such a system is now in the early stage of development. It has admirable environmental qualities, bonus attractions for industrial processors, potentially very significant international benefits, and a strong touch of glamour and adventure to appeal to both public officials and the public. The proposal to construct in space, using lunar materials, a series of solar power stations, and space industrial facilities, has gained considerable and usually favorable notice from industry, government agencies, elected officials and the general public in both the U.S. and several other countries. The literature on this proposal is rapidly growing. The basic concept can be found in several sources.[50–54] Both NASA and ERDA are working on the project; several private firms have invested money and manpower, and an international private citizens' group (the L-5 Society) very actively proselytizes on behalf of the system. The major product of the solar satellites would be electrical power beamed to the planet by microwave, but the engineering and economic studies which have been conducted to date suggest that within perhaps 30 years the power quantities available would be so large and the cost so low that hydrogen derived from the electrolysis of water could be used to replace hydrocarbon fuels. If hydrogen economy advocates were to fit their proposals into such a system and assist in gaining support for it, an eclectic electric-hydrogen economy would seem very feasible, since it would meet some of the conditions mentioned in sections 3.8. and 3.9.

Whatever system is used to generate hydrogen, transportation through a pipeline, perhaps both a new specially-designed line and an existing one formerly used for natural gas, would provide necessary data on transmission technology and on both absolute costs and costs in comparison to other energy transportation methods. If the hydrogen can then be used in an industrial application, or perhaps as a suburban electrical power source (fuel cell experimentation?), both safety questions and public acceptability should be tested.

Any model hydrogen economy system should include not only possibilities for technological development, but some political "angles" which could be designed in and used to good effect if the experiment proves successful. The model system would provide a baseline for both technical and social impact studies, and economic information of value would also be produced. Special-purpose applications, such as NASA's use of liquid hydrogen for rocket vehicles or fuel cells, may tell us something of technological value but are of little use in evaluating social impact, political reactions, or even environmental significance. The social scientist is unfortunately not permitted to manipulate societies, rerun elections to test the impact of changed campaign strategy, or perform other useful "laboratory" experiments. If a model hydrogen system could be designed in consultation with those who must evaluate its social impact and predict its future role in society, there would be considerable advantages for both the technology and the society.

REFERENCES

1. **Forrester, J. W.,** Counterintuitive nature of social systems, *Technol. Rev.,* 73, 53, 1971.
2. **North, R. C.,** Some paradoxes of war and peace, *Papers of Peace Sci. Soc. (Int.),* 25, 1, 1975.
3. **Meadows, D. H., Meadows, D. L., Randers, J., and Behrens, W. W., III,** *The Limits to Growth,* Universe Books, New York, 1972; Mesarovic, M. and Pestel, E., *Mankind at the Turning Point,* E. P. Dutton, New York, 1974.
4. **Bauer, R. A., Rosenbloom, R. S., and Sharp, L.,** *Second-Order Consequences: A Methodological Essay on the Impact of Technology,* MIT Press, Cambridge, Mass., 1969.
5. **Wolf, C. P., Ed.,** *Social Impact Assessment,* Environmental Design Research Association, Milwaukee, 1974.
6. **Soroos, M. S.,** Research note: behavioral science, forecasting, and the designing of alternative future worlds, *Int. Interactions,* 1, 269, 1974.
7. **Gilmore, J. S., Matthews, W. E., and Duff, M. K.,** *Hydrogen: Socio-environmental Impact,* University of Denver Research Institute, April, 1975.
8. **Seamans, R. C., Jr.,** *Chronicle of Higher Education,* Nov. 10, 1975, 10.
9. **Cook, E.,** The flow of energy in an industrial society, *Sci. Am.,* 224, 134, 1971.
10. Ford Foundation Energy Policy Project, *A Time to Choose,* Ballinger, Cambridge, Mass., 1974.
11. **Cole, H. S. D., Ed.,** *Models of Doom,* Universe Books, New York, 1973.
12. Congressional Research Service, for the Special Subcommittee on Investigations of the Committee on International Relations, Oil fields as military objectives: a feasibility study, 94th Cong., 1st Session, U.S. Government Printing Office, 1975.
13. Ford Foundation Energy Policy Project, *A Time to Choose,* Ballinger, Cambridge, Mass., 1974, chap. 5.
14. **Davis, D. H.,** *Energy Politics,* St. Martin's Press, New York, 1974.
15. Ford Foundation, Energy Policy Project, *A Time to Choose,* Ballinger, Cambridge, Mass., 1974, chap. 5.
16. **Truman, D. B.,** *The Governmental Process,* Knopf, New York, 1953.
17. **Madison, J., Hamilton, A., and Jay, J.,** *The Federalist,* Modern Library, New York, 1937, Papers 10, 51.
18. **Schattschneider, E. E.,** *The Semi-Sovereign People,* Holt, Rinehart & Winston, New York, 1960.
19. **Easton, D.,** *A Systems Analysis of Political Life,* John Wiley & Sons, New York, 1965.
20. **Starr, C.,** Social benefit vs. technological risk, *Science,* 165, 1232, 1969.
21. **Lindblom, C. E.,** *The Intelligence of Democracy,* The Free Press, New York, 1965.
22. **Bell, D.,** *The Coming of Post-Industrial Society,* Basic Books, New York, 1973.
23. **Edelman, M.,** *The Symbolic Uses of Politics,* University of Illinois Press, Urbana, 1964.
24. **Wildavsky, A.,** *Dixon-Yates: A Study in Power Politics,* Yale University Press, New Haven, 1962.
25. **Davis, D. H.,** *Energy Politics,* St. Martin's Press, New York, 1974, chap. 1.
26. **Hellman, R.,** *Government Competition in the Electric Utility Industry,* Praeger, New York, 1972.
27. **Escher, W. J. D.,** Prospects for liquid hydrogen fueled commercial aircraft, Escher Technology Associates, Report PR-37, 1973; Dyer, C. R., Sincoff, M. Z., and Cribbins, P. D., Eds., *The Energy Dilemma and Its Impact on Air Transportation,* NASA NGT 47-003-028, NASA-Langley Research Center, 1973.
28. **Davis, D. H.,** *Energy Politics,* St. Martin's Press, New York, 1974, 9.
29. **Enloe, C. H.,** *The Politics of Pollution in a Comparative Perspective,* McKay, New York, 1975.
30. **Woll, P.,** *Public Policy,* Winthrop, Cambridge, Mass., 1974, chap. 1.
31. **Hammond, A. L., Metz, W. D., and Maugh, T. H., III,** *Energy and the Future,* American Association for the Advancement of Science, Washington, D.C., 1973, chap. 16.
32. International City Managers Association, *The Municipal Yearbook 1972,* I.C.M.A., Washington, D.C., 1972, 106.
33. **Accum, F.,** *Practical Treatise on Gas-Light,* Vol. 3, Davies, Michael, and Hudson, London, 1816, 180.
34. **Weinberg, M.,** Can technology replace social engineering?, in Mesthene, F. G., Ed., *Technology and Social Change,* Bobbs-Merrill, Indianapolis, 1967.
35. **Key, V. O.,** *Public Opinion and American Democracy,* Knopf, New York, 1964.
36. **Enloe, C. H.,** *The Politics of Pollution in a Comparative Perspective,* McKay, New York, 1975, chap. 5; see also **Allison, G. T.,** *Essence of Decision,* Little, Brown, Boston, 1971.
37. **Halperin, M. H.,** *Bureaucratic Politics and Foreign Policy,* Brookings Institution, Washington, D.C., 1974, 99.
38. International City Managers Association, *The Municipal Yearbook 1972,* I.C.M.A., Washington, D.C., 1972, 105.
39. **Downs, A.,** Up and down with ecology – the 'issue-attention cycle', *Public Interest,* 28, 38, 1972.

40. ERDA 76-1, A National Plan for Energy Research, Development and Demonstration: Creating Energy Choices for the Future, U.S. Energy Research and Development Administration, Washington, D.C., 1976.
41. **Huntington, S. P.,** Strategic planning and the political process, *Foreign Affairs,* 38, 285, 1960.
42. **Plass, H. J., Jr.,** How might the hydrogen economy affect our resources and environment, in Veziroglu, T. N., Ed., *Hydrogen Energy,* Part B, Plenum, New York, 1975, 1157.
43. **Flattau, E. and Stansbury, J.,** It takes energy to produce energy: the net's the thing, *Washington Monthly,* 20, March 1974; Odum, H., *Environment, Power and Society,* Wiley-Interscience, New York, 1971.
44. **Epstein, E. M.,** *The Corporation in American Politics,* Prentice-Hall, Englewood Cliffs, N.J., 1969.
45. Congressional Quarterly, *1974 CQ Almanac,* Vol. 30, Congressional Quarterly Inc., Washington, D.C., 1975, 753.
46. **Fuller, J. G.,** *We Almost Lost Detroit,* Readers Digest Press, New York, 1975.
47. **Savage, R. L. et al.,** *A Hydrogenergy Carrier,* Vol. 2, NASA NGT 44-005-114, Houston, 1973, chap. 7.
48. **Woll, P.,** Public Policy, Winthrop, Cambridge, Mass., 1974, chap. 8.
49. Ford Foundation, Energy Policy Project, *A Time to Choose,* Ballinger, Cambridge, Mass., 1974, chap. 7.
50. **O'Neill, G. K.,** The Colonization of Space, *Phys. Today,* 27, 32, 1974.
51. **O'Neill, G. K.,** Space Colonies and Energy Supply to the Earth, *Science,* 190, 943, 1975.
52. **O'Neill, G. K.,** *Space Manufacturing from Non-terrestrial Materials,* NASA-Ames Laboratory and Office of Aeronautics and Space Technology, Mountain View, Cal., 1976.
53. U.S. Senate Subcommittee on Aerospace Technology and National Needs of the Committee on Aeronautical and Space Sciences, Hearings on Solar Power from Satellites, 94th Congress, 2d Session, January 19 and 21, 1976.
54. **O'Neill, G. K.,** *The High Frontier,* Morrow, New York, 1977.

Chapter 4

Legal Aspects of Hydrogen

Chapter 4

LEGAL ASPECTS OF HYDROGEN

Thomas C. Cady

TABLE OF CONTENTS

4.1. INTRODUCTION

The current energy crisis is not simply a technical/economic problem requiring technical/economic solutions. It is also, among many other things, a legal problem requiring a legal solution. The interaction of technology and law may be imagined as the equivalent of a totally mirrored room. An image is reflected and rereflected 1000-fold. Any technological change precipitates legal change, and vice versa and so on. The introduction of hydrogen as a major fuel in the American energy system will significantly affect and, in turn, will be significantly affected by the legal system.

4.1.1. Purpose, Organization, and Methodology

It is the purpose of this chapter to investigate the legal aspects of a transition to greater production, distribution, and use of hydrogen in the American Energy system. There are two major legal aspects of hydrogen: domestic and international. The domestic legal aspect section begins with a brief introduction presenting the concept of energy law, then surveys the past legal order and recent trends, and projects those into the future of the years 2000 to 2025. A summary and conclusion end the domestic aspect. The international legal aspect section is similarly organized. Finally the chapter ends with a general conclusion.

Predicting the legal future is hazardous if not impossible. There are no hard data or firm formulas. Law moves and lawyers think interstitially — on a case-to-case or statute-to-amendment basis. Lawyers are loath to state positively even what the law is, let alone predict what the law will be in the future. Having thus disclaimed any ability to do so, we begin our quest to determine the legal aspects of hydrogen in the years 2000 to 2025.

4.2. DOMESTIC LEGAL ASPECTS OF HYDROGEN

4.2.1. Introduction

Because of a providence of natural and human resources not duplicated elsewhere in the world, Americans have believed most profoundly in growth. The assumption was that energy sources were virtually limitless and could, without end, fuel that growth. Quite naturally, the law reflected the basic premise of abundance and gave rise to a very distinctive system of law — energy law.

Energy law does not exist as a distinct area of law. It is an abstraction designed to bring together for present analytical purposes separate legal areas. Energy law contains two main parts — property law and regulatory law. The contours of energy law may perhaps be best understood by the use of a unifying concept: it is the concept of shared regulation. Shared, in this context, means a sharing of power among the federal, state, and local governments and private units. Regulation means a rule or order guiding action issued by a public (government) or private executive. Authority is scattered between public and private decision makers and among federal, state, and local responsibility. Despite this diversity, two dominant features have stood out. First, because of a strong feeling in our polity in favor of the reserved police power of the states, they, rather than the federal government, have had the primary and overwhelming responsibility. Second, because of equally strong adherence to economic doctrines of *laissez faire*, most energy decisions were fundamentally private.

Recent legal developments, however, indicate that there is a definite trend away from private decision making towards public decision making and, as well, a definite trend away from local and state responsibility towards federal government control. The hydrogen energy system of the future will probably be privately owned in name only. In short, it will be centralized.

4.2.2. Property Law

4.2.2.1. Introduction

American property law is deeply rooted in English feudalistic concepts concerning land. Land was the basis of wealth and thus required special protection. However, the word "property" does not mean only a thing, but rather denotes a legal relation between a person and a thing. The thing may be an object having physical existence or it may be any kind of an intangible such as a patent right or a choice in action. Thus, property is really a right to assert control over something of value. The control right is not unlimited and is protected by several constitutional provisions. The interplay is well expressed in a recent Supreme Court opinion:[1]

> Subject to limitations imposed by the common law of nuisance and zoning restrictions, the owner of real property has the right to develop his land to his own economic advantage. As land continues to become more scarce, and as land use planning constantly becomes more sophisticated, the needs and the opportunities for unforeseen uses of specific parcels of real estate continually increase.

4.2.2.2. Nuisance

The common law nuisance action is an ancient method of controlling pollution. A nuisance is a unreasonable interference with the use and enjoyment of land.[2] A nuisance may have a limited impact and be classified as a private nuisance, to be litigated by a limited number of property owners, or a nuisance may affect a large number of the public and be classified as a public nuisance to be litigated by a public prosecutor. Remedies available were either an injunction ordering the polluter to cease operations altogether or to stop the pollution by corrective devices or an award of damages.

Nuisance law has had little development in the recent past. Perhaps the most significant recent addition to nuisance law is the famous *Del Webb/Sun City* case.[3] There, the defendant owned and operated a rendering plant far outside the Phoenix city limits. During the 1950s and 1960s, the city experienced a dramatic population growth. Plaintiff then developed its famous retirement city, Sun City, which completely encircled defendant's plant. Plaintiff sued, urging the defendant's operation be totally stopped. Defendant countered with the historical argument that Plaintiff had moved to the nuisance and, hence, had assumed the risk of any interference with property. The Arizona Supreme Court, in a landmark decision, combined the two remedies. The rendering plant was required to shut down and move, but the plaintiff was required to share in the costs of the move. While praised in the literature, the case has not found favor in subsequent decisions.

The future of nuisance law is very difficult to predict. No trend has emerged. This area of law appears to be fairly stable. It may be likely that, as energy decisions become more difficult, the critical term — unreasonable (as the test of whether or not the interference is a nuisance) — will change. The change may be in the direction of a more relaxed standard in favor of energy. On the other hand, with increasing environmental concerns and demands for internalization of social overhead costs, the slightest pollution may be held to be unreasonable and hence result in a nuisance liability award against the polluter.

4.2.2.3. Zoning

One recently developed limitation on the ownership and use of land is zoning. While zoning theoretically could be statewide or regional, it is usually at a city level. Zoning is defined as governmental (rather than private market) regulation of land use without compensation to the owners of land. It was not until 1926 in the leading case of *Euclid v. Ambler Realty Company*[4] that the U.S. Supreme Court first upheld zoning as a constitutional exercise of governmental power. Ambler Realty had sued the city of Euclid arguing that the city zoning ordinance restricted Ambler in the development of its land. Such restriction, Ambler argued, violated the Fourteenth Amendment, depriving it of its property without due process of law. The Court noted that local governments retain a great amount of power to legislate for the public good, and zoning was a clearly constitutional exercise of that power. The power to zone, however, was not unlimited. A zoning restriction would be unconstitutional if it were "clearly arbitrary and unreasonable, having no relation to the public health, safety, morals, or general welfare,"[5] or did not have a method by which an individual property owner could apply for an exception or variances.[6]

A difference exists between the adoption of a city-wide comprehensive zoning plan and the subsequent decisions allowing or denying a specific use of a specific piece of property. Adoptions are subject to a limited review by the courts and may be attacked only on the basis that the comprehensive zoning plan was "clearly arbitrary and unreasonable". On the other hand, subsequent decisions concerning specific uses on specific pieces of property are subject to a more expanded review by the courts. Due process requires, in such cases, that the interested parties be given a reasonable opportunity to be heard before an impartial and qualified decision maker applying applicable and reasonable standards. Zoning, by its very nature, interferes significantly with the owner's use of property. A mere diminution in the market value or a frustration of a proposed use are insufficient to invalidate a zoning ordinance or to entitle the owner to a variance or rezoning. However, a zoning action may be of such a nature as to severely depreciate the value of the land. Such action would clearly violate due process. It may perhaps amount to a taking of private property for public use without just compensation in violation of the Fifth Amendment of the Constitution.

Originally, zoning ordinances merely classified a city into various zones such as residential, commercial, and industrial with appropriate procedures for allowing exceptions and variances. Most recently, however, many cities have enacted so-called exclusionary zoning ordinances. The effect of such ordinances is to restrict severely or to exclude altogether some activity or structure or to set a certain density of inhabitants. The Supreme Court has just recently approved of such zoning in *Village of Belle Terre v. Boraas.*[7] There, the village restricted land use to one-family dwellings, thus prohibiting all lodging houses, boarding houses, fraternity houses, or multiple dwelling houses. The village thereby virtually foreclosed any immigration and growth. Nevertheless, the Supreme Court found that the power to zone to be extremely broad and[8]

A quiet place where yards are wide, people few, and motor vehicles restricted are legitimate guidelines in a land-use project addressed to family needs. This goal is a permissible one within *Berman v. Parker, supra.* The police power is not confined to elimination of filth, stench, and unhealthy places. It is ample to lay out zones where family values, youth values, and the blessings of quiet seclusion, and clean air make the area a sanctuary for people.

The most recent decisions of the Supreme Court on zoning reaffirm the broad power to zone, even to the extent of allowing voter ratification of a proposed zoning change. In sum, then, the recent trend in zoning decisions is a reaffirmation of the basics announced in the very first Supreme Court decision.

Unlike the situation of the recent past, most municipalities are neither isolated nor independent. Zoning decisions in a tiny municipality may affect several cities, a whole region, a state, or even several states. The present system of fragmented zoning decisions does not adequately serve regional, state, and national needs. National and regional energy needs could be frustrated by the decision of a single small village "not here". Since the power of a municipality to zone is a power delegated to it by the state, we can expect that the zoning power in the future will be highly centralized with very large regional bodies (two or three to a state) or a single state commission making all of the major zoning decisions.

4.2.2.4. Land Use Planning

Many municipalities have extended the broad power of zoning to a much more complicated and sophisticated form of zoning called land use planning. Land use planning differs from zoning in that zoning merely involves regulation of the various types of uses permitted within various areas of a city, whereas land use planning involves not only regulation of uses but an overall goal for the community and a plan for severely controlled (or even zero) growth. There is really no past to this fast developing area of property law. The state of the law consists only of recent trends.

Only two very recent cases are on point. In the first, *Golden v. Planning Board of Town of Ramapo,*[9] the highest court of New York, the Court of Appeals, approved a controlled growth plan. There, the town of Ramapo experienced a substantial increase in population with the resulting strain on existing municipal facilities and services (sewage, drainage, parks, schools, roads, fire and police protection) and financial resources. After a comprehensive study, the city amended its zoning ordinance in 1969. It adopted a phased growth plan covering a period of 18 years designed to eliminate premature subdivision and urban sprawl by projecting a detailed capital improvements plan for *maximum* growth of existing land. Several plaintiffs (a property owner, a developer, and builders) sued, alleging the Ramapo plan was an unconstitutional confiscation of their property. The trial court found the plan constitutional and the appellate division found it unconstitutional, but the Court of Appeals on a 5 to 2 split vote finally held the plan to be constitutional. The Court held that the plan Ramapo sought[10]

. . . not to freeze population at present levels but to maximize growth by the efficient use of land, and in so doing testify to this community's continuing role in population assimilation. In sum, Ramapo asks not that it be left alone, but only that it be allowed to prevent the kind of deterioration that has transformed well-ordered and thriving residential communities into blighted ghettos with attendant hazards to health, security and social stability — a danger not without substantial basis in fact. . . . In sum, where it is clear that the existing physical and financial resources of the community are inadequate to furnish the essential services and facilities which a substantial increase in population requires, there is a rational basis for "phased growth" and hence, the challenged ordinance is not violative of the Federal and State Constitutions . . .

The most recent and most famous controlled growth land use plan was held to be constitutional in *Construction Industry Association of Sonoma County v. City of Petaluma.*[11] The city of Petaluma, located about 40 mi north of San Francisco, had experienced relatively controlled growth during the 1950s and 1960s. In 1970 and 1971, the city had a population spurt of almost 25%. Alarmed by the accelerated and uncontrolled growth, the city council adopted a development policy in June 1971. The preamble stated,[11a] "In order to protect its small town character and surrounding open spaces, it shall be the policy of the City to control its future rate and distribution of growth." In 1972, the council adopted several resolutions with collectively were called the Petaluma Plan. The plan was limited to a 5-year period from 1972 to 1977, fixed a housing development rate, a 200-ft-wide greenbelt around the city to serve as a boundary for urban expansion, and an innovative Residential Development Control System which provided allocation procedures and criteria for the award of housing development permits on an intricate point system. Two landowners and an association of builders challenged the plan claiming it was unconstitutional. The U.S. District Court found the plan was an unconstitutional violation of the right to travel. The city appealed. The U.S. Court of Appeals for the Ninth Circuit reversed the lower court. It held that the plaintiffs had no standing to assert the plan was a violation of the constitutional right of unknown third parties (not before the court) to travel. The Court of Appeals went further. It also held that the plan did not violate due process because the concept of the public welfare involved in zoning and land use planning was sufficiently broad to uphold the desire of the city to preserve its small town character, its open spaces, and its low density of population and to grow at an orderly and deliberate pace. The Court carefully pointed out that its decision was not a permanent endorsement of the plan. Finally, the Court held that the plan was not an impermissible burden on interstate commerce.

The significance of the *Petaluma* decision cannot be underestimated: it significantly broadened the power of municipalities to control growth far beyond that approved by the New York Court in *Ramapo.* In *Ramapo,* the plan was based on maximum but phased growth, while in *Petaluma,* the plan was based on a strict limitation of population growth. *Petaluma* was unique in that it is the first case to approve of an outright numerical limitation on population growth. No court prior to the *Petaluma* decision had approved such broadened land use planning power. Surely the approval of the Petaluma Plan will prompt other communities to adopt similar plans. The effect will be of devastating impact on future energy decisions. So long as communities have such broad powers to limit growth and to make restrictive land use plans, where in the U.S. could an energy development project take place? The result would be a severely Balkanized U.S. "freezing to death in the dark" in small, limited-growth communities.

The recent trends will not continue. The probable solution seems to have been foreseen in *Petaluma.* The power of local communities to plan for land use is not an inherent right of power vested in the locality. It is a delegated power of the state. As soon as local concerns tend to frustrate regional or statewide needs, the states will surely recapture the power and establish regional or state agencies to make large-scale land use decisions. Such a result is clearly indicated in *Petaluma.*

> If the present system of delegated zoning power does not effectively serve the state interest in furthering the general welfare of the region or entire state, it is the state legislature's and not the federal courts' role to intervene and adjust the system. As stated *supra,* the federal court is not a super zoning board and should not be called on to mark the point at which legitimate local interests in promoting the welfare of the community are outweighed by legitimate regional interests . . . [12]
>
> Our decision upholding the Plan as not in violation of the appellees' due process rights should not be read as a permanent endorsement of the Plan. In a few years the City itself for good reason may abandon the Plan or the state may decide to alter its laws delegating its zoning power to the local authorities; or to meet legitimate regional needs, regional zoning authorties may be established. . . . However, the federal court is not the proper forum for resolving these problems. The controversy stirred up by the present litigation, as indicated by the number and variety of *amici* on each side, and the complex economic, political and social factors involved in this case are compelling evidence that resolution of the important housing and environmental issues raised here is exclusively the domain of the legislature.[13]

The state authority might be modeled on the New York example. Because of a complex of regulatory and environmental concerns over a period of years, no new energy plant of any kind (atomic, coal, or oil) had been constructed in the state of New York. The State created a State Board of Electric Generating Siting and the Environment. The Board includes representative of the State Public Service Commission, the Department of Environmental Conservation, the Department of Commerce, the Department of Health, and a resident in the area of the proposed site.

4.2.2.5. Eminent Domain

All land in the U.S. is owned; there is no such thing as unowned land. Land ownership is divided into three kinds: private, Indian, and public. Private ownership includes individuals, partnerships, and corporations. Indian ownership includes ownership by individuals and tribes exclusive of federally owned land used by Indians. The chart below summarizes the subject as of 1969:[14]

Classes	Land acres (millions)	Percent
Private land	1317	58.2
Indian land	50	2.2
Public land	897	39.6
Federal land	763	33.7
State land	114	5.0
County and municipal land	20	0.9
Totals	2264	100

The predominance of private ownership of property in the U.S. raises problems of acquisition of property for any expansion of any aspect (new mines, plants, pipelines, etc.) of hydrogen. Of necessity, most of the land acquired will be private. Proposals that hydrogen production plants be sited on land or that major new undergroud pipeline distribution systems be constructed will require that the lands so acquired must be condemned under the power of eminent domain.

The right of eminent domain is an inherent power of the federal government. All private and public property is subject to the superior power of eminent domain of the federal government. The states also have the power of eminent domain but do not have the power to condemn federal lands. All eminent domain power is controlled by the Fifth Amendment to the Constituion which requires that property can be taken only for a public use and that just compensation must be paid.

Four questions arise: (1) What is a "taking"? (2) What is "property"? (3) What is a "public use"? (4) What is "just compensation"? A "taking" has been defined to have occurred "when inroads are made upon the owner's use of it to an extent that, as between private parties, a servitude has been acquired either by an agreement or in course of time."[15] Thus, flights by aircraft directly over[16] or even over adjacent property,[17] which, because of noise vibration and danger, render the property unusable have been deemed to have been the taking of an easement. "Property" may be defined as any legal property right including the land itself,[18] improvements,[19] leases,[20] franchises,[21] letters patent,[22] Liens,[23] air,[24] or water.[25] A "public use" is whatever the legislature declares to be a public use. The legislative declaration is an exclusive perogative of the legislature and is not subject to judicial second-guessing.[26] "Just compensation" is the fair market value of the property taken. Fair market value is not an absolute standard nor an exclusive method of valuation. Fair market value derives its meaning from the technical concepts of property law — what a willing buyer in an open market would pay in cash to a willing seller — and basic equitable principles of fairness.[27]

While no significant recent trends could be found in this apparently quietly settled area of property law, it is anticipated that the power of eminent domain will be extensively used to acquire property needed for the introduction of hydrogen into the changing energy system. The American energy system basic concepts of eminent domain, being etched in constitutional writ, will remain essentially unchanged. Hydrogen will be expensive.

4.2.2.6. Special Problems

4.2.2.6.1. Land Ownership Disputes Among and Between the States and the U.S.

A land boundary between states is generally a simple matter of the location of a beginning point and a course of directions and distances. A water boundary between states is a bit more complicated. It is usually the high or low water mark on either bank. It may also be the "meander line" (as the water slowly changes course) or the geographic center or, finally, the "thalweg" (the middle of the deepest or most navigable channel). The lateral marine boundary separating coastal states may prove to be the most troublesome. It may be any one of the above[28] or as specified in the Convention on the Territorial Sea and Contiguous Zone. The Supreme Court has recently praised the Convention as being most helpful because "the comprehensiveness of the Convention provides answers to many of the lesser problems related to coastlines which, absent the Convention, would be most troublesome."[29]

Quite early in the history of the U.S., the Supreme Court adopted the "Equal Footing Doctrine". It requires that all new states admitted to the Union have the same and equal sovereignity, jurisdiction, power, economic, and property rights as the older states. That is to say, every new state was to have the same rights as the 13 original states. The Equal Footing Doctrine has had a tremendous impact on the property rights of the states to submerged lands. In *Pollard's Lessee v. Hagan,*[30] the Supreme Court held that since the original states upon formation of the Union had reserved to themselves ownership of all of their submerged lands under internal waters, the new states, under the Equal Footing Doctrine, are entitled to retain the property rights to such submerged lands. Then in a series of most unfortunate decisions between 1947 to 1950, the Supreme Court refused to extend the inland water rule to submerged lands off the ocean[31] and gulf coasts.[32] Finally, in *United States v. Texas,*[33] the Equal Footing Doctrine was applied in vengeful reverse. Although Texas admittedly owned submerged lands into the Gulf of Mexico, upon its admission to the Union and under the Equal Footing Doctrine it automatically surrendered ownership to those submerged lands to the U.S. The magnitude of the economic stakes involved — billions of dollars — was staggering.[34]

> States and their grantees have expended millions of dollrs to build piers, breakwaters, jetties, and other structures, to install sewage-disposal systms and to fill in beaches and reclaim lands. During the past two decades California, Louisiana, and Texas have been leasing substantial portions of the lands in question for oil, gas, and mineral development. California commenced such leasing in 1921 and Texas in 1926. Other States, including Washington, Florida, Mississippi, North Carolina, and Maryland, have made leases for like purposes. States have levied and collected taxes upon interests in and improvements on these lands. . . . Senator Holland placed the figure at "billions of dollars of invested money."

Congress reversed the decision in 1953 by enacting the Submerged Lands Act.[35] The Act had four major aspects:

(1) Relinquishes to the states the entire interest of the U.S. in all lands beneath navigable waters within state boundaries
(2) Defines that area in terms of state boundaries "as they existed at the time [a] State became a member of the Union, or as heretofore approved by the congress," not extending, however, seaward from the coast of any state more than 3 marine leagues in the Gulf of Mexico or more than 3 geographical mi in the Atlantic and Pacific Oceans
(3) Confirms to each state a seaward boundary of 3 geographical mi, without "questioning or in any manner prejudicing the existence of any State's seaward boundary beyond three geographical miles if it was so provided by its constitution or laws prior to or at the time such State became a member of the Union, or if it has been heretofore approved by Congress"
(4) For purposes of commerce, navigation, national defense, and international affairs, reserves to the U.S. all constitutional powers of regulation and control over the areas within which the proprietary interests of the states are recognized and retains in the U.S. all rights in submerged lands laying beyond those areas to the seaward limits of the continental shelf.

The Act also confirmed the ownership rights of the Great Lake states in owning their adjacent submerged lands to the international border with Canada. Significantly, the Act reversed to the U.S. the exclusive right to the "use, development, improvement or control" of all water power or the use of water for the production of power. In a tumultous decision[36] involving all of the Gulf of Mexico coastal states (Texas, Louisiana, Mississippi, Alabama, and Florida), the Supreme Court confirmed the rights of the Gulf Coast states to the lands, minerals, and resources underlying the Gulf of Mexico.

In summary, now the coastal states bordering on the Atlantic and Pacific Oceans own the submerged lands off their coasts for a distance of 3 geographical miles. The coastal states bordering on the Gulf of Mexico, with the exception of Texas, own the submerged lands off their coasts for a distance of 3 geographic miles. Texas, because of special historical reasions, owns its submerged lands for a distance of 3 marine leagues or about 10½ geographic miles.

Subsequent decisions have confirmed the seward rights of other states in the Pacific[37] and Atlantic[38] Oceans. The most recent decision by the Supreme Court on the Equal Footing Doctrine was in *Oregon ex rel. State Land Board v. Corvallis Sand & Gravel Co.*[39] There, the state brought an action to determine the ownership of certain lands underlying a navigable but not an interstate boundary. The Supreme Court strongly reaffirmed the Equal Footing Doctrine and held that almost all submerged lands under internal waters are to be controlled by state property law. In so holding, the Supreme Court overruled one of its decisions that was only 3 years old. It was strongly suggested that, except for the application of federal property law to fixing the boundaries of a riverbed acquired by a state at the time of its admission to the Union, federal property law ends and thereafter the submerged land is subject to the property law of the state.

Almost all of the property disputes among the states and between the states and the U.S. will be finally settled in the future. Proposals for the siting of hydrogen production facilities offshore should be made with caution. Under the Submerged Lands Act, the U.S. reserved to itself the ownership of the water power and the use of water for the production of power. If the facility were sited on submerged lands owned by a state, it would be subject to all the laws of that state as well as state taxation. Since the U.S., under the Submerged Lands Act, reserved to itself ownership of the water power and the use of water for the production of power, the facility sited on U.S. property would enjoy an economic advantage.

4.2.2.6.2. Indian Lands

Approximately 50 million acres of land are owned by individual Indians or Indian tribes. The heaviest concentration of ownership is in the plains and mountain states of South Dakota, Oklahoma, Montana, Wyoming, New Mexico, Utah, Arizona, Washington, and Nevada. The acreages run from a high of 20,034,000 acres in Arizona to a low of 1,153,000 acres in Nevada.[40] Traditional Indian beliefs concerning land are very special. For example, the Hopi believe:[41]

> Hopi tradition regards the arid, inhospitable land of northeastern Arizona as the center of the universe, destined to be the environment of a pure people who would retain their sacred traditions. The people of the high mesas have a great reverence of their *Hopi Tusuqua*, or land. To the Hopi, the land is more than a given amount of earth; it is part of their identity, their very being. The Hopi believes that nature and god are one and that physical objects are imbued with spirits which can be appealed to for aid.

Similarly, for the Navajo:[42]

> Traditional Navajo beliefs reflect a reverence for their land. The Navajo look upon the earth as a material figure who provides for the physical and spiritual needs of her children. This world view is also reflected in the structure of Navajo society. For example Navajos trace their descent through the mother for inheritance of land use rights.

The Hopi and Navajo Reservations have economically significant known reserves of coal, oil, and uranium. Because of depressed economic conditions, some coal leases have been sold, but this has sparked sharp intra- and intertribal controversies.[43] Further, acquisition or development of Indian land for production, transportation, or storage of hydrogen will face strong religious barriers.

4.2.2.7. Water Law

Energy development requires water. Water must be used for processing, cooling, cleaning, and transportation. Hydrogen energy development will involve increased uses of water in one or more of the above-mentioned areas. Increased uses of water require acquisition of water and acquisition of water involves the law of ownership of water as governed by property law. There are two systems of water law in the United States.

In the eastern U.S. (the Mississippi Valley and eastward) the property law of water is the law of riparian rights. Under the eastern theory, water is owned by the owner of the land through which the water flows. All land owners on the water have equality of ownership of the water. All land owners, and hence water owners, have the right to reasonable use of the water. Neither the location of the land above or below other owners, nor the size of the land, nor the priority of ownership or use make any difference. Reasonable use means that each riparian owner of the water stands in a position of equality and each may make reasonable use of the water. Reasonable use also means that each riparian owner may use all the water he needs but he must insure the continued size, flow, and purity of the water. Obviously, the two prior statements are in conflict during water shortages and thus during such times all riparian owners must share the shortage equally.

In the seventeen western states of the U.S., the property law of water is the prior appropriation doctrine. Water is owned by the person who has made a priority of beneficial use of the water. Water ownership is granted by its use by time and purpose — "first come, first served". The prior appropriation owner may continue to use and own such amounts of water for such purpose in perpetuity without regard to subsequent demands of others for the water no matter, generally, the reasonableness of their needs.

Under both systems of water law, water is owned. Owned water simply cannot be taken. A taking must be taken according to the property principles of due process and eminent domain. Any proposal for hydrogen energy development involving any water use involves water law and, therefore, property law. In short, water use rights will have to be bought and, hence, hydrogen will be expensive.

4.2.3. Regulatory Law

4.2.3.1. Introduction

The other significant domestic legal aspect of hydrogen is regulatory law. It includes economic, safety, antitrust, and environmental regulation. Regulation of the American energy system is an example of the basic concept of shared regulation. Responsibility was widely shared among public and private decision makers and among levels of government. Recent trends in regulatory law (much more so than in property law) indicate a definite trend away from local and private decision making to centralized public decision making or federal preemption of regulatory law.

4.2.3.2. Economic Regulation

Economic regulation of the American energy system at the federal level is done in the public interest. It is generally defined as being twofold: first, to insure an efficient and dependable energy system to meet the needs of a growing national economy and, second, to serve the needs of consumers by insuring abundant, low-cost, safe energy. These purposes, however, may conflict. Regulation of a company or industry to insure its viability and growth in an expanding economy requires promotion of the specific company or industry. The extent to which the company or industry is promoted may result in diminished concern for the needs of the consumers. Similarly, the reverse is true: the regulatory agency may see as its mandate the protection of the consumer with resulting harm to the company or industry. Two examples will serve to illustrate this problem.

The Atomic Energy Act was amended in 1954 for the specific purpose of encouraging private enterprise to assist in the development and utilization of atomic energy for peaceful purposes. In the recent case of *Northern States Power Company v. Minnesota,*[44] a Federal Appeals Court noted that Congressional objectives in the 1954 amendments "evince [a] legislative design to foster and encourage the development, use and control of atomic energy so as to make the maximum contribution to the general welfare and to increase the standard of living."[45]

This is a pretty heady mandate and, as can be expected, consumer views tended to be lost. For example, in *Crowther v. Seaborg,*[46] an objection was made to an extremely hazardous underground nuclear explosion to test the feasibility of nuclear stimulation of gases. Residents and landowners complained that the Atomic Energy Commission (AEC) had approved the test without full knowledge of the total impact of the test. While the court sympathized with the fears of residents and landowners, it supported the counterargument of the AEC that its job was to stimulate the use and development of atomic energy.

On the other hand, the regulatory agency may see its obligation to the consumer as primary. The result is abundant, low-cost energy but a severely damaged industry. Such was the case with the Federal Power Commission (FPC) and the natural gas industry. The FPC set such low rates for natural gas that an artificially high demand for the low-cost fuel was created. As markets expanded, exploration incentives were reduced. It as been asserted that this action contributed to the current shortages in natural gas. In its 1972 report, the FPC itself admitted that "the Commission sought to rectify past regulatory practice so as to provide independent producers with the required financial incentives for investing the huge amounts of risk capital without which the exploration and development of gas is impossible . . . " and that "the prospective curtailments during the winter heating season of 1972—1973 offered harsh proof that capital formation during the past decade had failed to provide financial incentives adequate to spur domestic gas producers to explore and develop gas resources."[47]

Congress, however, has relied upon the states in many instances to supplement the economic regulatory power of the federal government. Even when federal authority has been exercised, the usual pattern is a dual system of federal and state authority. State authority is most pervasive in the coal mining industry where comprehensive safety codes, siting requirements, and pollution standards abound. State residual authority in the petroleum industry is likewise quite comprehensive. A well-known illustration is the powerful Texas Railroad Commission. We find in the areas of natural gas and electricity that the states retain the residual power of economic regulation, mostly by fixing consumer prices. Perhaps the most highly state-regulated aspect of the energy system is of electric power generation, distribution, and sale.

Economic regulation of the American energy system seems to be tending to federal preemption. In *Otter Tail Power Company v. United States*,[48] the Supreme Court was presented with the question of the refusal of a power company to deal with certain city corporations that wanted to provide their own service to residents by purchasing, at wholesale prices, electrical energy produced by the company. While the case concerned the anticompetitive effect of the refusal of the power company to deal, the Court noted that the Federal Power Act, although quite comprehensive, still allowed dual governance of economic decisions between the public sector (federal government) and the private sector (private owners). Thus, private ownership is maintained and commercial relationships are governed, in the first instance, by the private business judgment of the company. This result is so despite the fact that the company was a highly regulated natural monopoly. In other words, there is still a zone of freedom allowed by a regulatory act where private decision making is generally unregulated; however, that area seems to be shrinking rapidly.

Similarly, in the many instances where federal regulation is not exclusive but rather shared with the states, the trend to federal preemption grows. Such an issue was recently involved in *Federal Power Commission v. Louisiana Power & Light Company*.[49] That case presented the fundamental question of whether or not the FPC was empowered by the Federal Power Act to take certain regulatory action over the company. The company argued that the FPC had no jurisdiction. The Supreme Court, construing the Natural Gas Act of 1938, noted that the FPC had been granted broad powers "to protect consumers against exploitation at the hands of natural gas monopolies." To that end Congress "meant to create a comprehensive and effective regulatory scheme" of dual state and federal authority. The Court also noted that although federal jurisdiction was not to be exclusive, FPC regulation was to be broadly complementary to that reserved to the states so that there would be no "gaps" for private interests to subvert the public welfare. Thus, the question became: "which jurisdiction should fill the gap?" To this question the Court answered that[50]

> when a dispute arises over whether a given transaction is within the scope of federal or state regulatory authority, we are not inclined to approach the problem negatively, thus raising the possibility that a "no man's land" will be created. That is to say in a borderline case where congressional authority is not explicit, we must ask whether state authority can practically regulate the given area and if we find it cannot, then we are impelled to decide that federal authority governs.

Noting that there is inevitably a conflict between producing states and consuming states (which could create contradictory actions, regulations, and rules that could not possibly be equitably resolved by the courts), the Court felt that a uniform federal regulation was desirable. The Court, therefore, concluded that the matter in question was indeed within the jurisdiction of the FPC. The important point of the case is that competition for energy among the states gives rise to the diversity of state resolutions, thus impelling the Court to find that the area is federally regulated. Regulation in the national interest is ensured, but the regulatory power of the states diminishes.

For a number of years, spokesmen for the energy industry have noted the need for a national energy policy and coordination among the various federal and state regulatory agencies. Among President Carter's first legislative assignments was the creation of the Department of Energy on August 4, 1977.[51] The Department consolidated many of the regulatory functions scattered throughout the federal structure. Thus, it may be that the federal and state regulatory structures will be, in the future, consolidated into this or a similar, single federal cabinet-level department. Surely, as the demand for energy increases and the supplies decrease, this agency will be given more and more functions of research, development, funding, and regulation of the entire energy picture.

4.2.3.3. Safety Regulation

Safety regulation of the American energy system at the federal level has been minimal. There are, however, several areas in which the federal government plays a dominant safety role. The first area is that of wholly federally owned energy facilities. As of June 1972, the federal government owned about 42% (24.1 million kW) of all the hydroelectric generating capacity in the contiguous U.S.[52] Since the Tennessee Valley Authority is a federally owned and created energy facility, safety is primarily a federal responsibility. Where energy takes on an interstate character, the federal government will assume dominant regulatory responsibility. Of all the oil pipelines in the U.S., 85.1% are regulated by the Interstate Commerce Commission.[53] The FPC has been authorized to regulate the interstate aspects of electrical and natural gas transmission and, thus, has primary safety responsibility. The AEC, since 1946, has held pervasive authority over atomic energy. The enabling legislation is replete with Congressional concern for protection of the public health and safety. While no case yet held that the federal act has ousted state authority for safety decision, there are some small indications that such may be the result.[54]

By a large margin, the primary and overwhelming responsibility for safety regulation in the U.S. has been in the states. This is due to a very strong belief that, in our federal system, the states have the reserved power to protect and promote the health, safety, and general welfare of its citizenry.[55] Every state has enacted broad-ranging safety codes, but weak programs and insufficient enforcement led to the Federal Occupational Safety and Health Act of 1970.[56] State energy codes are concerned mostly with job safety, equipment, and the construction and use of materials in and around the home. Housing codes, however, are generally adoptions of privately generated codes. Another area of state energy safety code responsibility is the rather limited professional and trade codes, for example, the licensing of electricians. Trends indicate that there will be increased licensing and professional requirements for trades involved in the U.S. energy system.

Another area of state responsibility for safety is that of remedies for accidents. Every state has enacted a workmen's compensation system. These systems provide a low-level, no-fault recovery for workers injured as a result of a work-connected accident. For persons not workers (i.e., the public) injured by an energy-related accident, the states also provide a common law tort remedy. These laws generally require proof of fault such as negligence or recklessness on the part of an agent of an energy supplier. Strict liability or a no-fault liability may be applied to extrahazardous activities involving energy.

Of particular significance in the area of state safety responsibility is a recent first impression case decided by the U.S. Court of Appeals for the Eighth Circuit — *Northern States Power Company v. Minnesota.*[57] There, the central question posed was whether the U.S. government had the sole authority under the doctrine of preemption to regulate radioactive waste released from nuclear power plants to the exclusion of any state authority. The court concluded that Congress had vested the AEC with authority to resolve the proper balance between desired industrial progress and adequate health and safety standards and that only through the application and enforcement of nationally uniform standards, promulgated by a national agency, could the dual objectives of industrial progress and adequate health and safety be met. The court noted that should the states be allowed to impose stricter standards on the level of radioactive waste, they might conceivably be so overprotective as to unnecessarily stultify industrial development. This result was reached despite the fact that the conventional approach had been to recognize the legitimate interests of the states to protect the health and safety of its citizens. Previously, courts had refused to find federal preemption over state health and safety laws absent clear and unmistakable expression. Preemption, on the other hand, was much more readily found in cases not involving public health and safety. Here, without precedent, it is suggested that (1) where national interests are involved in a search for energy, (2) where there is thus a collision with the traditional power residing in the states to protect the health and safety of its citizens, and (3) where Congress has not expressly preempted state law, then the courts may find preemption in general Congressional legislation affecting energy use. This is truly startling.

Two recent federal safety acts — the Federal Coal Mine Health and Safety Act of 1969 and the Occupational Safety and Health Act of 1970 (OSHA)[58] — have substantially altered the traditional federal-state roles in safety legislation. The 1969 Coal Mine Act introduced two brand-new features into the U.S. energy system. First, it wrote into federal law a comprehensive safety code for that energy industry. Second, it introduced the federal government into the workman's compensation system (thought heretofore to be an exclusive state prerogative). Note that this legislation vitally affects the U.S. energy system, since coal is now the main source of energy for electric generation plants and is projected to become increasingly important as supplies of other fossil fuels diminish. As revolutionary as the 1969 Coal Mine Act was, the 1970 OSHA is positively radical. The act potentially covers all employers (with some exceptions for governmental employers otherwise treated) whose business affects interstate commerce. (As a footnote, virtually any business can be found to affect interstate commerce.) The Secretary of Labor is given the power to rule safety by administrative standards. While Congress did not preempt the states from retaining any authority over occupational health and safety, the political dynamics are such that the "OSHA eventually will be the primary force in the employee-safety field preempting contrary state action" for virtually every industry in America.[58a] This is surely a radical first in American safety legislation.

Thus, it may be seen from a survey of federal-state relations in safety regulation that a trend is developing: (1) increasing federal responsibility in both safety codes and workmen's compensation and (2) the increased federal regulation of safety will be to the preemption of state authority to set higher or different standards.

It is significant to note that the *Northern States Power Company* case mentioned above involved atomic energy — one of the proposed prime energy sources to be used in the production of hydrogen. Should hydrogen enter the energy system, we can expect comprehensive federal regulation of hydrogen and a comprehensive safety code. Also, should a conflict result between federal and state authority over safety and health conditions connected with the use of hydrogen, we can expect that the courts, even in the absence of legislative expression, will find that Congress has preempted state authority.

4.2.3.4. Antitrust

The economic doctrine of *laissez faire* profoundly influenced the American faith in private free enterprise and our form of economic organization. Competition was emphasized. The basic antitrust act — the Sherman Antitrust Act — was passed by Congress in 1890. In 1914 the Sherman Act was supplemented by the Federal Trade Commission (FTC) Act and the Clayton Act. These three acts were the core of American antitrust policy. The policy was to ensure a competitive economy by limiting concentration of economic power.

In one of the first antitrust cases to reach the U.S. Supreme Court — *Standard Oil Co. of New Jersey v. United States*[59] — the Court declared that the Rule of Reason was the general rule of construction of the act. The Court thus gained the discretion to decide whether conduct is significantly and unreasonably anticompetitive. Over the years, the acts have been given varying interpretation and emphasis, but generally the Rule of Reason has controlled.

During the 1960s, a growing list of oil companies started acquiring interests in competing fuels. The result was intensive horizontal integration in the fossil fuel industry. For example, by 1970, of the 25 largest oil companies, 18 had involvements in oil shale, 11 in coal, 18 in uranium, and 7 in tar sands. This trend created what is now commonly called the "Energy Company".[60] The Energy Company phenomenon has led to a renewed application of the antitrust laws to control the concentration of energy resources. It should be noted, however, that it was long ago recognized that there must be many natural monopolies. While quite contrary to basic notions of free enterprise competition, natural monopolies are very much a legal part of the American scene. Even so, a natural monopoly may be subject to antitrust action. In the recent case of *Gulf States Utilities Company v. Federal Power Commission*[61] the Supreme Court held that the broad regulatory authority of the FPC over electric utility companies included the responsibility for considering possible anticompetitive aspects in authorizing a utility security issue. Responsibility was necessary, the Court felt, because it provided the first line of defense against anticompetitive practices which might later become subject to antitrust proceedings. Also, in another 1973 Supreme Court case, *Otter Tail Power Company v. United States,*[62] the Court held that even though a utility may be quite highly controlled by a regulatory agency, it is nevertheless subject to independent scrutiny under the antitrust laws. Repeals of the antitrust laws, by implication from a regulatory scheme, are to be strongly disfavored and have only been found, said the Supreme Court, in cases of plain repugnancy between the antitrust laws and the regulatory provisions.

On another antitrust front, a recent suit filed by the Florida Attorney General charged 15 major U.S. oil companies with conspiring to violate the antitrust laws by creating a nationwide fuel shortage. If successful, the suit could be the "biggest trustbusting attempt since the breakup of Standard Oil in the 1900's."[63] The suit alleged that the oil companies have engaged in an illegal monopoly and in unreasonable restraints of interstate commerce. It also alleged that the oil companies conspired to reduce competition, raise, fix, and stabilize prices, and excluded nonmajor oil companies from significant aspects (e.g., refining and exploration) of the industry. The relief sought is that the companies should be forced out of crude oil exploration and production. In yet another antitrust case, in July 1973, the FTC accused the eight largest U.S. oil companies of monopolistic refining and marketing practices. Paralleling the Florida suit, the FTC charged that the big oil firms have controlled the market from oil well to gas pump for the last 25 years, making it almost impossible for new companies to enter the refining business. The FTC action, in filing a formal antitrust complaint, was most unusual since it often issues a proposed informal complaint, thus quietly giving the parties involved a chance to settle the matter out of court.[64] As of September 1978, the case has dragged on for 5 years, but the FTC has been persistent. In the latest move in the massive case, a federal administrative law judge rejected most of the companies' objection to a FTC subpoena for certain corporate documents for 12 years through 1977. The FTC is seeking an order requiring divestiture by the companies (Exxon, Texaco, Gulf, Mobil, Standard Oil of California, Standard Oil of Indiana, Shell, and Atlantic Richfield) of 40 to 60% of their refining capacities in certain markets and divestiture of certain pipelines.[64a]

All this renewed interest in application of the antitrust law is sure to have a profound effect. The horizontally and/or vertically integrated energy company may be broken up into a severely atomized system. With capital investment costs approaching the limits of all but the largest company, it is anticipated that a vigorous application of the antitrust law to reduce company size may significantly hinder private entry into the hydrogen economy.

4.2.3.5. Environmental Law

One of the most dramatic and recent developments of the American legal scene has been the growth of environmental law. It has been suggested that environmental activism began in the 1950s, when the focus was on atmospheric nuclear testing, then moved to pesticides in the 1960s,[65] and finally, in the 1970s, to national policy. On January 1, 1970, the National Environmental Policy Act (NEPA) became law.[66] It has been called the most important piece of environmental legislation ever written and will surely have a major impact on energy decisions in the future.

The act declares that it is the policy of the federal government "to use all practicable means and measures, including financial and technical assistance, in a manner calculated to foster and promote the general welfare, to create and maintain conditions under which man and nature can exist in productive harmony, and fulfill the social, economic, and other requirements of present and future generations of Americans."[67]

To implement that policy, the act directs that, to the fullest extent possible, *all* policies, regulations, and laws shall be interpreted and administered in accord with the purposes of the act. Also required is the now famous NEPA environmental impact statement. The statement must survey the following five items:[68]

- The environmental impact of the proposed action
- Any adverse environmental effects which cannot be avoided should the proposal be implemented
- Alternatives to the proposed action
- The relationship between local short-term uses of man's environment and the maintenance and enhancement of long-term productivity
- Any irreversible and irretrievable commitments of resources which would be involved in the proposed action should it be implemented.

The significance of NEPA can be seen by a comparison of two cases — one pre-NEPA and the other post-NEPA — both involving the AEC. In 1969 (prior to NEPA), a U.S. Appeals Court found that the AEC need not consider the thermal pollution impact of a proposed nuclear power plant.[69] Only 2 years later, after the enactment of NEPA, another court found that the regulations of the AEC were inadequate because they failed to consider fully environmental factors, including thermal pollution.[70]

It must be emphasized again that NEPA requires *all* agencies of the federal government (including courts and agencies involved in the energy system) to interpret and administer *all* policies, regulations, and laws in accord with the environmental purposes of the act. Supplementing NEPA are various federal acts designed to protect specific areas of the environment. The environmental attractiveness of hydrogen as a fuel and/or energy carrier would fit in well with the NEPA requirement. Production methods, however, may present significant problems.

Every state and locality has power to protect the environment. Because of the great diversity, no attempt will be made to catalog or organize the laws; rather, some highlights will be mentioned. Perhaps the most significant recent development has been the inclusion of environmental protection in state constitutions. Florida, Illinois, Michigan, New York, Pennsylvania, Rhode Island, and Virginia have recently adopted constitutional provisions concerning the environment. Some of these enact constitutionally based provisions giving the citizen a right to a clean environment.[71]

The basis for state and local regulation of the environment is the traditional idea of police power possessed by state and local governing bodies. The conventional rule has been that these powers were not to be superseded by an act unless there was a clear Congressional intent. The trend, however, seems to be that Congress (with the assistance of the courts) will find, in ever-increasing areas of environmental control, a need for a federalized environmental law.

Preemption of state authority for environmental protection appears to be increasing. The trend is indicated by two recent cases. In the first case, *Northern States Power Company v. Minnesota,* a federal court held, in an important first impression case, that the AEC regulations had preempted the state of Minnesota from any regulation of the levels of radioactive effluents discharged from nuclear generation plants. Also noteworthy is *City of Burbank v. Lockheed Air Terminal.* The Supreme Court noted that Congress, in enacting the Noise Control Act of 1972 and the 1972 Amendments to the Federal Aviation Act, involved the EPA in a comprehensive scheme of federal control of the aircraft noise problem. The Court concluded that Congress had preempted state and local control over aircraft noise. This conclusion was reached despite the fact that there was no express provision of preemption in any of the acts. It was the pervasive nature of the scheme of federal regulation of aircraft noise that led the court to find preemption. It may be that these two cases are portents of the future wherein there will be increasing federal preemption in areas previously thought to be exclusively the domain of the state for environmental protection.

4.2.4. Summary and Conclusions

What conclusions can be drawn from this brief survey of the current American energy system? The trend (both judicially and legislatively) is to federal preemption. As problems of energy use and supply become more critical, many functions will be taken away from the states by the federal government. This is indicated by the decision of the Supreme Court in *FPC v. Louisiana Power & Light Company* where we found the U.S. Supreme Court saying that a federal regulatory agency, in effect, abhors an energy system vaccuum. It is also illustrated by *Northern States Power Company v. Minnesota* where the eighth Circuit held, in a first impression decision, that the traditional power of a state under the TenthAmendment to protect and promote the health, safety, and general welfare of its citizens must give way to a uniform national policy for the development and utilization of nuclear energy. The most emphatic conclusion that can be made, however, is that the American energy system, will be almost entirely federalized by the 21st century. It will be privately owned in name only.

4.3. INTERNATIONAL LEGAL ASPECT OF HYDROGEN

4.3.1. Introduction

The current energy crisis has profound implications in international law. As the world moves from a fossil fuel based economy to new forms of energy, the movement will involve the oceans and economic policies.[72] More than any other area of law previously discussed, international law is in the midst of furious and radical change.[73] In the last two decades, international law concepts that had remained relatively unchanged for hundreds of years have been completely rewritten. Future developments in the remainder of this century will find a wholly new law of the sea and of international economics.

4.3.2. The International Law of the Sea

Up until the end of World War II, the international law of the sea was clearly settled. The basic principle was freedom of the seas. The principle was first announced by the famous Dutch jurist, Hugo Grotius, in his 1604 landmark work, *Mare Librum.* The assumption was that the sheer vastness of the seas of the world and the staggering quantity of their resources (an abundance of space and resources) could accommodate all navigational uses and resource exploitation without limit.[74] Thus, the seas and their resources were owned by no one, free for use and exploitation by all mankind. A coastal state owned territory to the water's edge but all the seas belonged to no one and to everyone.

Even as Grotius wrote, however, many coastal states had asserted territorial jurisdiction over the seas within a measured distance of their coasts. The purposes were defensive and for the preferential treatment of coastal fishing and trade. In 1610, Holland, then in the midst of a fishing dispute with the British, was probably the first coastal state to assert the "Cannon Shot Rule."[75] The rule was premised on the idea that a coastal state could exercise jurisdiction over the seas within the range of its coastal armament. Whatever the range of a coastal cannon was in the 17th or 18th century, customary international practice settled on a 3-mi limit to the territorial sea.[76] It also became customary in international law that a coastal state could exercise protective and preventive law enforcement jurisdiction (but not ownership) over a band of the high seas contiguous to its territorial sea for the purpose of preventing and punishing violation of important domestic laws concerning customs and smuggling. Since 1790, the U.S. had asserted jurisdiction over a 12-mi contiguous zone for custom purposes. Again, traditional international practice settled on a limit of 12 mi from the shore.[77] Thus, up until World War II, the state of the international law of the seas can be easily summarized as in Figure 1.

By the end of World War II, the international law of the sea began to unravel. In 1945, President Truman, on behalf of the U.S., asserted by proclamation the theretofore-unheard-of idea that the U.S. had jurisdiction, control, and ownership of the sea bed and subsoil of the "continental shelf".[78] The proclamation also asserted that the U.S. had the right to establish exclusive fishing zones reserved for Americans only and the right to establish conservation zones in other areas of the sea above the shelf. The continental shelf, first mentioned in 1898 by a geographer,[79] was a technical term used to describe the gently sloping plain which underlies the seas extending from the adjacent land mass to the sharp drop-off, which is then the ocean depths. The continental shelf may vary in width from only 1 mi to, in the Gulf of

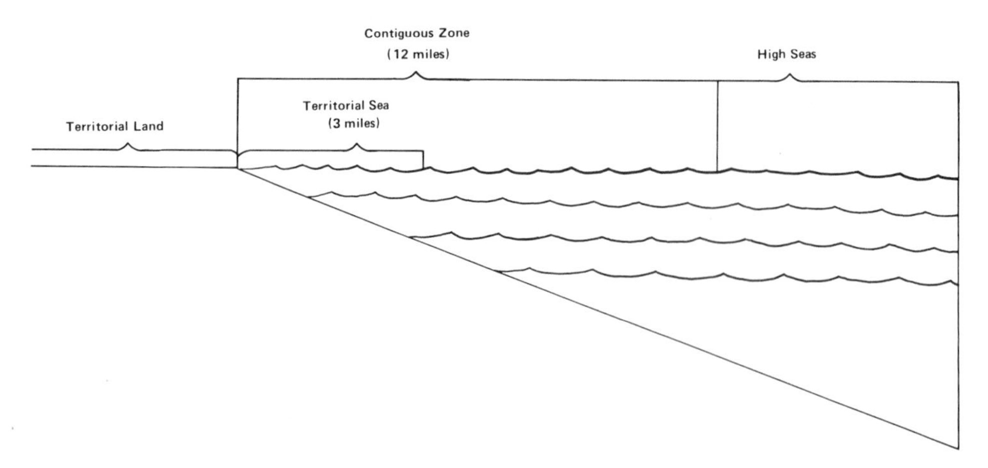

FIGURE 1. The state of the international law of the seas — 1945.

Mexico, as much as 200 mi.[80] The average width is about 30 mi.[81] The Truman Proclamation not only created a new concept in international law, but it shattered the old order.

Many coastal states followed the American example, and some even expanded upon it. Within a few years after the Truman Proclamation, several coastal states began to claim an expanded territorial sea up to 12 mi. The most breath-taking claims were those made in the 1952 Declaration of Santiago where Chile, Ecuador, and Peru (later joined by Costa Rica) announced that they had created conservation zones extending for 200 mi off their respective coasts and that each had the right to exclusive control of such zones.[82] Amid the welter of claims and denials and the shrinkage of the high sea, the first United Nations Conference on the Law of Sea was convened in 1958 in Geneva. That conference produced four Conventions:

- The Convention on the Territorial Sea and the Contiguous Zone
- The Convention on the Continental Shelf
- The Convention on the High Seas
- The Convention on Fishing and Conservation of the Living Resources of the High Seas

The first three conventions became effective but the fourth has, to date, not entered into force.

- The Territorial Sea: The concept of a Territorial Sea was confirmed. It is a part of the territory of the coastal state and, as such, it and all the resources in and below it are owned by the coastal state. No agreement was reached on its breadth.[83]
- The Contiguous Zone: The Contiguous Zone concept was accepted. A coastal state could exercise protective and preventive jurisdiction (but no ownership) over a band of the high sea contiguous to its territorial sea for its customs, fiscal, immigration, or sanitary regulations only. The breadth of the Contiguous Zone could not extend beyond 12 mi from the shoreline.[84]
- The Continental Shelf: The Continental Shelf concept was approved. The coastal state was given the exclusive ownership rights of the sea bed and subsoil and resources on or below the continental shelf. The breadth of the area was limited to a depth of 200 mi or, beyond that limit, to where a depth technology allowed exploitation of the resources. The superadjacent waters remained a part of high sea.
- The High Sea: The High Sea regime was also settled. The high seas were defined as all parts of the sea that were not territorial sea. The high seas were open to all nations and subject to ownership by none.

Thus, the state of the international law of the sea, in 1958, can be summarized as in Figure 2.

Because the participants of the 1958 Geneva Conference could not agree on several important issues (principally, the breadth of the territorial sea and fishing rights) a second Conference on the Law of the Sea was called in 1960. No agreement could be reached on any issue.

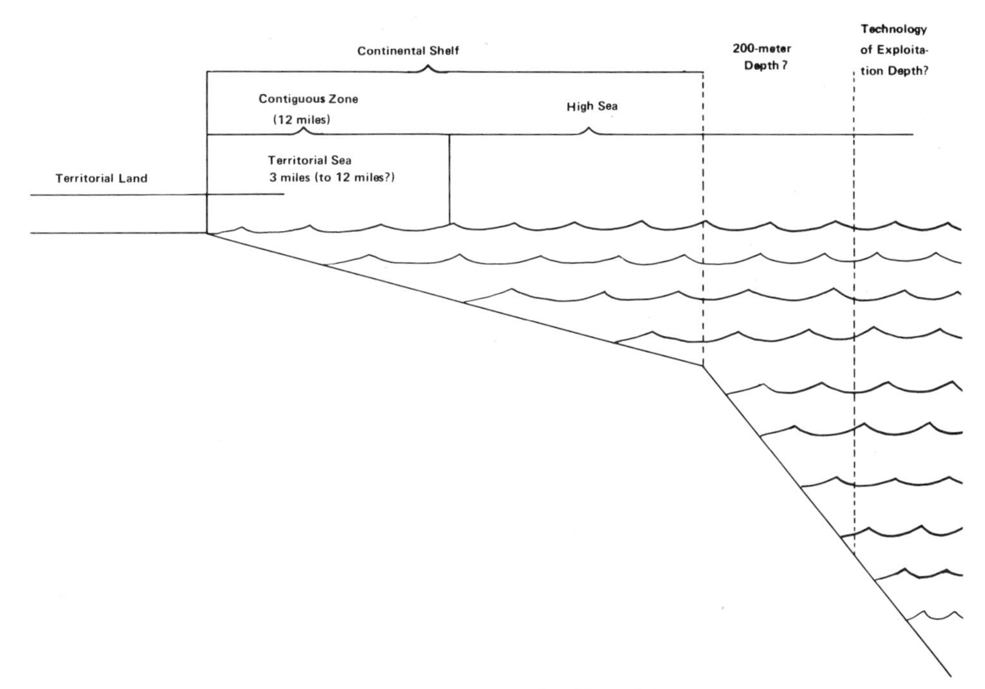

FIGURE 2. The state of the international law of the seas — 1958.

Since the breadth of the high seas was not agreed upon, a majority of the coastal states of the world began to extend their territorial sea. As of 1971, the greatest number had picked 12 mi, while nine coastal states (Argentina, Brazil, Chile, Ecuador, El Salvador, Panama, Peru, Sierra Leone, and Uruguay) extended their respective territorial seas to 200 mi.[85] Further, many coastal states began to establish exclusive "conservation zones" designed to insure preference to domestic fishers. Many international incidents followed, the most famous being "The Cod War" between Iceland and England.[86] Finally, another new issue emerged concerning the mining of mineral-rich nodules on the floors of the high seas. In 1970, the United Nations adopted a resolution known as the "Declaration of Principles" which provided that the sea bed, the ocean floor, and its resources beyond the limits of national jurisdiction are the common heritage of mankind, not subject to any claim or exploitation until an appropriate international regulatory agency is set up.[87] In 1975, the U.S. Department of State received letters asserting claims of ownership or requests for exclusive deep sea mining or petroleum development rights. All were denied on the grounds that international law did not prevent exploration.[88]

In hopes of resolving these issues, a third United Nations Conference on the Law of the Sea began meeting in 1974. In 1976, a working text known as the "Revised Single Negotiating Text", generally regarded as the consensus of the participants, was produced. It specified as follows:[89]

1. A maximum 12-mi limit for the territorial sea, over which the coastal state will have sovereignty, subject to a right of innocent passage, with some elaboration of the rules of innocent passage
2. Unimpeded passage of straits used for international navigation for all vessels and aircraft
3. A 200-mi economic zone in which the coastal state exercises sovereign rights over the exploration, exploitation, conservation, and management of living and nonliving resources and in which all states continue to enjoy freedoms, in particular of navigation and overflight and other uses related to navigation and communication; coastal state sovereign rights over the exploration and exploitation of the resources of the seabed and subsoil of the continental margin where it extends beyond 200 mi, coupled with a duty to contribute some international payments in respect of mineral production in the area of the margin beyond 200 mi
4. Comprehensive coastal state control of all drilling and of all economic installations in the economic zone
5. Some adjustment and modernization of the regime of the high seas, for example, the recognition of the special interest and responsibility of the state of origin for anadromous species of fish and new rules with re-

spect to control of unauthorized broadcasting and cooperation in the suppression of illicit traffic in narcotics
6. An elaboration of a concept of island nations as archipelagic states which includes a precise definition of a new category of archipelagic waters and a regime of unimpeded passage through archipelagic sealanes and air routes that traverse the archipelago
7. International standards to prevent and control marine pollution and limited coastal state enforcement rights with respect to vessel-source pollution
8. Specified coastal state and flag state rights and duties with respect to scientific research in the economic zone and on the continental shelf and general provisions regarding international cooperation in marine scientific research and transfer of marine technology
9. An international regime and machinery to deal with the exploration and exploitation of seabed resources beyond the limits of national jurisdiction (that is, beyond the economic zone or continental margin)
10. A system for peaceful third-party settlement of disputes regarding the interpretation or application of the Convention which have not been resolved by negotiation or other agreed procedures.

Some significant points should be noted. The territorial sea was expanded to 12 mi. The contiguous zone and continental zone disappeared and were combined into a 200-mi economic zone. The economic zone is a revolutionary concept in the international law of the sea. It is a hybrid zone comprising elements of ownership found in territorial sea and elements of nonownership found in the high sea concept. The intermingling of those two concepts can be seen in the two most important articles concerning the economic zone:[90]

Article 44

1. In an area beyond and adjacent to its territorial sea, described as the exclusive economic zone, the coastal State has:
   a. Sovereign rights for the purpose of exploring and exploiting, conserving and managing the natural resources, whether living or nonliving, of the bed and subsoil and the superjacent waters;
   b. Exclusive right and jurisdiction with regard to the establishment and use of artificial islands, installations and structures;
   c. Exclusive jurisdiction with regard to:
      i. Other activities for the economic exploitation and exploration of the zone, such as the production of energy from the water, currents and winds; and
      ii. Scientific research.
   d. Jurisdiction with regard to the preservation of the marine environment, including pollution control and abatement;
   e. Other rights and duties provided for in the present Convention.
2. In exercising its rights and performing its duties under the present Convention in the exclusive economic zone, the coastal State shall have due regard to the rights and duties of other States.
3. The rights set out in this article with respect to the bed and subsoil shall be exercised in accordance with Chapter IV.

Article 46

1. In the exclusive economic zone, all States, whether coastal or land-locked, enjoy, subject to the relevant provisions of the present Convention, the freedoms of navigation and overflight and of the laying of submarine cables and pipelines and other internationally lawful uses of the sea related navigation and communication.
2. Articles 77 and 103 and other pertinent rules of international law apply to the exclusive economic zone in so far as they are not incompatible with this Chapter.
3. In exercising their rights and performing their duties under the present Convention in the exclusive economic zone, States shall have due regard to the rights and duties of the coastal State and shall comply with the laws and regulations enacted by the coastal State in conformity with this Chapter and other rules of international law.

Approximately 85% of the evolving international law of the sea is agreed upon. About 10 to 15 important issues require resolution.[91] A sixth session of the third Conference began in New York on May 23, 1977 and continued until July 15, 1977.[92] A completed treaty is expected in late 1978 or early 1979. The state of the international law of the sea can be tentatively summarized as in Figure 3.

It has been suggested that territorial claims to the high sea follow the technology to exploit the resources of the high sea. Complete territorial claims appear to have been limited for the time being at 12 mi, while partial territorial claims are limited at 200 mi. The current negotiations reflect again the ancient tensions between the conflicting principles of *Mare Liberum* and *Mare Clausim*. Recent trends clearly indicate a victory for the latter and rush of "creeping nationalization" of ocean space. Many proposals concerning the production, storage, and distribution of hydrogen focus on the ocean. As the high seas are becoming increasingly less free, such proposals could raise serious territorial disputes.

4.3.3. The International Economic System

The old international economic system was based on the liberal economic principles of free

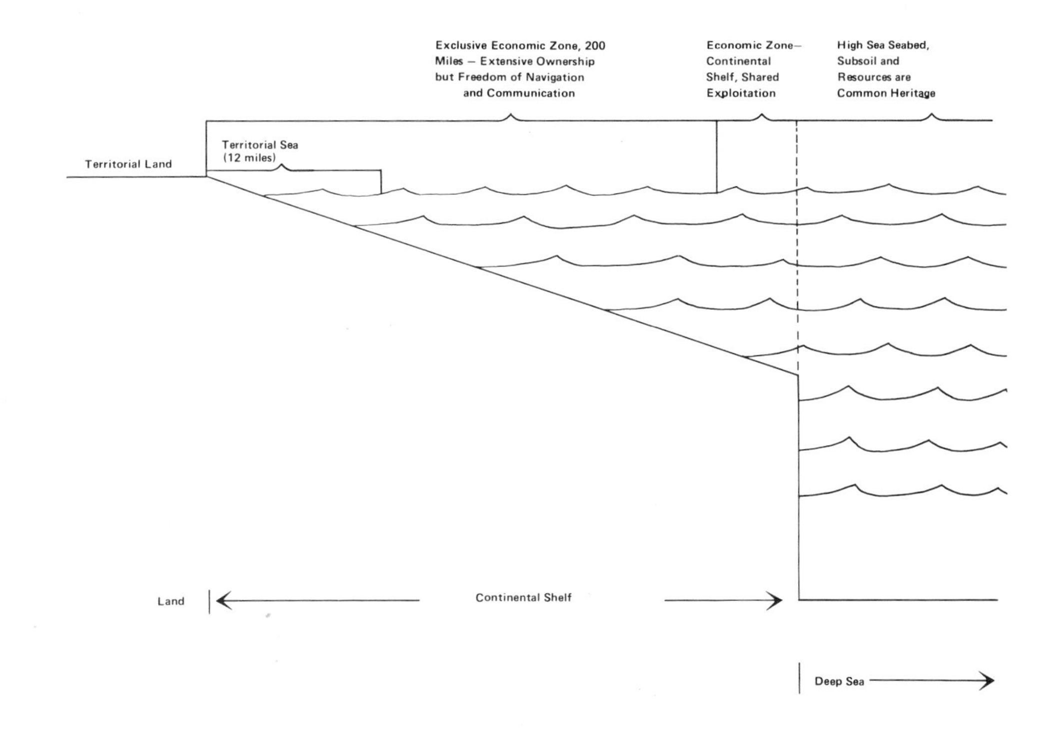

FIGURE 3. The state of the international law of the seas — 1978.

trade in free markets. The flow of capital, goods, and investments and price levels were set by the free market mechanism. Goods and capital were to flow freely on a nondiscriminatory or "most-favored-nation" basis. The objective of the most-favored-nation treatment was that private interests in A country were given the benefit in B country of the best economic opportunity given by B country to any alien's goods or capital, and vice versa. Multinational corporations were allowed to invest and operate but were subject to nondiscriminatory regulation and supervision. Private foreign investment was encouraged, but nationalization was allowed only if the taking was for a public purpose and just compensation was promptly paid. Because of a tangle of political, military, and economic concerns, governments always occupied a leading role in world trade. Critical to an orderly system was stability. Stability was provided by a complex network of bilateral treaties, agreements, practices, private contracts, and a dominant nation providing leadership. England from about 1870 to World War I was the leading stabilizer of the world. The U.S. then became the world leader for the 30-year period from the mid-1930s to the 1960s. Soon after the conclusion of World War II, the model for the postwar international economic system was set in the General Agreement on Tariffs and Trade (GATT). It provided for a "worldwide system of liberal trading, setting down procedures for lowering tariffs on a multilateral basis, providing for freedom of investment, limiting restrictive business practices, and establishing machinery for handling necessary exceptions and adjustments."[93] Many international and regional arrangements and agreements followed. Generally accepted international law supported the system. The system functioned quite well until the late 1960s and early 1970s.

The postwar years also saw the demise of the old colonial system. The new and developing countries of the world organized themselves into the Group of 77 in the United Nations Conference on Trade and Development in 1964. Subsequently, within the Group of 77, the Arab oil-rich states formed the Organization of Petroleum Exporting Countries (OPEC). The dramatic success of the OPEC oil embargo in December 1973 (quadrupling the price of oil virtually overnight) shattered the old order and provided the model for a new proposal as to how a new international economic system should be ordered.

The Group of 77 (now since grown to over 100) was able to push through the "Declaration on the Establishment of a New International Economic Order". It was adopted by the United Nations General Assembly without a vote, by consensus, on May 1, 1974. The Resolution began by noting that the most significant development during the last decades was the independence of nations from colonial rule, but that vestiges of colonialism remained which inhibited full emancipation. The benefits of technological progress were not distributed equitably, since the developing countries which account for 70% of the world population account for only 30% of the world income, and the gap was growing. The old economic order was responsible for the disequilibrium and must be replaced by the New International Economic Order (NIED). The NIED should include the following principles:

1. Accelerated development of all the developing countries
2. Full permanent sovereignty of every nation over its natural resources and all economic activities including the unalienable right to nationalization. (an implied rejection by silence of the requirement of just compensation for nationalization)
3. The right of former colonies to just compensation for past exploitation and depletion of or damages to the natural and other resources of former colonies (a radical reversal of the just compensation concept)
4. Full regulation and control of transnational corporations
5. Special aid to developing countries which were still under colonial domination
6. Price indexing of the products of the developing countries (a rejection of the free market mechanism in setting prices)
7. Preferential and nonreciprocal treatment for developing nations (a rejection of the most-favored-nation principle)
8. The transfer of financial resources to developing countries

4.4. CONCLUSION

The basic assumption of the American energy system was that energy sources were virtually limitless. Past and present law reflected that idea. The law of energy abundance prevailed. The old law essentially fostered maximum production, maximum distribution, and maximum consumption. Find it, drill it or dig it out, and then use it up. The law was ebullient, loose, market oriented — a crazy-quilt of contradictions, cross purposes, omissions, overlap, and duplication. In short, it was a law of exploitation well suited for an essentially *laissez faire* domestic and international economy.

As current, intermittent energy shortages grow into future chronic shortages, the law of the American energy system will change. The new law will be a law of energy scarcity. The new law will essentially foster limitation and conservation. It will be rigid, tight, need oriented — a single, comprehensive, federalized, internationalized, bureaucratic-administrative system of law. In short, it will be a law of allocation well suited for an essentially managed economy. Many of these future developments are now just trends. Great technological and social changes are in store for the U.S. The law is flexible and can be readily adjusted; it can be adopted quite well into a hydrogen economy.

REFERENCES

1. *City of Eastlake v. Forest City Enterprises, Inc.*, 426 U.S. 668, 681, 1976, (Stevens, J., dissenting).
2. **Prosser, W.**, *Law of Torts*, 4th ed., 1971, 86.
3. *Spur Industries, Inc. v. Del E. Webb Co.*, 108 Ariz. 178, 494 P.2d 700, 1972.
4. *Euclid v. Amber Realty Co.*, 272 U.S. 365, 1926.
5. *Euclid v. Amber Realty Co.*, 272 U.S. 395, 1926.
6. *Euclid v. Amber Realty Co.*, 272 U.S. 365, 1926.
7. 416 U.S. 1, 1974.
8. *Village of Belle Terre v. Boraas*, 416 U.S. 1, 9, 1974.
9. 30 N.Y.2d 359, N.Y.S.2d 138, 285 N.E.2d 291, appeal dismissed, 409 U.S. 1003, 1972.
10. 30 N.Y.2d 379—383, 334 N.Y.S.2d 152—156, 285 N.E.2d 302, appeal dismissed, 409 U.S. 1003, 1972.

11. 522 F.2d 897, 9th Cir., 1975, reversing 375 F. Supp. 574, N.D. Calif., 1974.
11a. 375 F. Supp. 574, 576, 76D Cal. 1974, reversed 522 F2d 897, 1975.
12. 522 F.2d at 908, 9th Cir., 1975.
13. 522 F.2d at 909, n. 17, 9th Cir., 1975.
14. Statistical Abstract of the United States, 207, 1976.
15. *United States v. Dickinson,* 331 U.S. 745, 748, 1947.
16. *United States v. Causby,* 328 U.S. 256, 1946.
17. *Griggs v. Allegheny County,* 369 U.S. 84, 1962.
18. *United States v. Fuller,* 409 U.S. 488, 1973.
19. *Almota Farmers Elevator & Whse. Co. v. United States,* 409 U.S. 470, 1973.
20. *Alamo Land & Cattle Co., Inc. v. Arizona,* 424 U.S. 295, 1976.
21. *Monongahela Navigation Co. v. United States,* 148 U.S. 312, 1893.
22. *James v. Campbell,* 104 U.S. 356, 1882.
23. *Armstrong v. United States,* 364 U.S. 40, 1960.
24. *Griggs v. Allegheny County, supra.,* note 17.
25. *United States v. Virginia Electric & Power Co.,* 365 U.S. 624, 1961.
26. *United States ex rel. T.V.A. v. Welch,* 327 U.S. 546, 1946.
27. *United States v. Fuller,* 409 U.S. 488, 1973.
28. *New Hampshire v. Maine,* 426 U.S. 363, 1976.
29. *United States v. California,* 381 U.S. 139, 165, 1965.
30. 44 U.S. (3 Harw.) 212, 1845.
31. *United States v. California,* 332 U.S. 19, 1947.
32. *United States v. Louisiana,* 339 U.S. 699, 1950.
33. 339 U.S. 707, 1950.
34. *United States v. Louisiana,* 363 U.S. 1, 94 n. 17, (Black concurring in part and dissenting in part), 1959.
35. 43 U.S.C.A. 1301, 1953.
36. *United States v. Louisiana,* 363 U.S. 1, 1959.
37. *United States v. California,* 381 U.S. 139, 1965; United States v. Alaska, 422 U.S. 184, 1975.
38. United States v. Maine, 420 U.S. 515, 1975; United States v. Florida, 425 U.S. 791, 1976.
39. 97 S. Ct. 582, 1977.
40. Statistical Abstract of United States, 212, 1976.
41. Comment, Of Indians, Land and the Federal Goverment: The Navajo-Hopi Land Dispute, 1976 Ariz. St. L. J. 173, 174 — 175.
42. Comment, Of Indians, Land the Federal Government: The Navajo-Hopi Land Dispute, 1976 Ariz. St. L. J. 177.
43. Comment, Of Indians, Land and the Federal Government: The Navajo-Hopi Land Dispute, 1976 Ariz. St. L. J. nns 64, 101.
44. 447 F.2d 1143, 8th Cir., 1971, aff'd mem., 405 U.S. 1035, 1972.
45. 447 F.2d 1143, 8th Cir., 1971, aff'd mem., 405 U.S. 1153, 1972.
46. 312 F. Supp. 1205, D. Colo., 1970.
47. 1072 FPC Ann. Rep. 1.
48. 410 U.S. 366, 1973.
49. 406 U.S. 621, 1972.
50. 406 U.S. 631, 1972.
51. 42 USCA 7111, 1977.
52. 1972 FPC Ann. Rep. 1.
53. 86 ICC Ann. Rep. 133, Table 4, 1972.
54. Compare Northern States Power Company v. Minnesota, 447 F.2d 1143, 8th Cir., 1971, af'd mem., 405 U.S. 1035, 1972, with Crowther v. Seaborg, 312 F. Supp. 1205, D. Colo., 1970.
55. *Supra.,* note 44.
56. 29 U.S.C. § 651, 1970.
57. *Supra.,* note 44.
58. 30 U.S.C. 801, 1970; *supra.,* note 56.
58a. Comment OSHA: Employer Beware, *Houst. L. Rev.,* 10, 426, 437, 1973.
59. 221 U.S. 1, 1911.
60. **Netschert, B. C.,** The energy company: a monopoly trend in the energy markets, *Bull. At. Scientists,* 27(8), 13, 1971.
61. 411 U.S. 747, 1973.
62. *Supra.,* note 48.
63. *Houston Post,* p.15A, July 10, 1973.
64. *Houston Post,* p.1A, July 18, 1973.
64a. *Wall Street Journal,* p.14 column 1, September 11, 1978.
65. **Lueck,** Energy and the environment, G. E. Co. TEMPO, GE 72 TMP-57, 51, 1972.
66. 42 U.S.C. § 4321, 1970.
67. 42 U.S.C. § 4321, 1970.

68. *Supra.*, note 67.
69. *New Hampshire v. AEC*, 406 F.2d 170, 1st Cir., 1969, cert. denied, 395 U.S. 962, 1969.
70. *Calvert Cliff Coordinating Comm. v. AEC*, 449 F.2d 1109, D.C. Cir., 1971.
71. **Howard, A. E.**, State constitutions and the environment, *Va. L. Rev.*, 58, 193, 1972.
72. **Borgese, E. M.**, The new international economic order & the law of the sea, *San Diego L. Rev.*, 14, 584, 1977.
73. **Pardo, A.**, Foreward, law of the sea. IX, *San Diego L. Rev.*, 14, 507, 1977.
74. **Pardo, A.**, Foreward, law of the sea. IX, *San Diego L. Rev.*, 14, 507, 1977.
75. **Kent, H. S. K.**, The historical origins of the three-mile limit, *A.J.I.L.*, 48, 537, 1954.
76. **Von Glahn, G.**, *Law Among Nations*, 3rd ed., 1976, 309.
77. **Von Glahn, G.**, *Law Among Nations*, 3rd ed., 1976, 332.
78. Presidential Proclamation No. 2667, September 28, 1945, 10 Fed Reg. 12303.
79. **Von Glahn, G.**, *Law Among Nations*, 3rd ed., 1976, 317.
80. *United States v. Louisiana*, 363 U.S. 1, 6 n. 3, 1959.
81. **Von Glahn, G.**, *Law Among Nations*, 3rd ed., 1976, 317.
82. **Von Glahn, G.**, *Law Among Nations*, 3rd ed., 1976, 336.
83. **Von Glahn, G.**, *Law Among Nations*, 3rd ed., 1976, 309.
84. Restatement, (Second) of Foreign Relations Law of U.S. § 21, 1965.
85. **Von Glahn, G.**, *Law Among Nations*, 3rd ed., 1976, 311.
86. **Von Glahn, G.**, *Law Among Nations*, 3rd ed., 1976, 334.
87. U.N.G.A. Res. 2749 (XXV), U.N. Doc. A/8097, 1970.
88. Digest of United States Practice in International Law, 427, 1975.
89. **Oxman, B. H.**, The third United Nations conference on the law of the sea: the 1976 New York sessions, *A.J.I.L.*, 71, 247, 1977.
90. **Clingan, T. A.**, Emerging law of the sea: the economic zone dilemma, *San Diego L. Rev.*, 14, 531, 1977.
91. **More, J. H.**, Introduction: next steps toward a law of the sea in the common interest, *San Diego L. Rev.*, 14, 523, 1977.
92. N.Y. Times, *p.21, June 12, 1977.*
93. **Kindleberger, C. B.**, U.S. foreign economic policy, 1776—1976, *For. Affairs*, 55, 395, 409, 1977.
94. *Int. Legis. Maritime*, 13, 715, 1974.

Unit Conversions, Physical Constants, and Symbols

UNIT CONVERSIONS, PHYSICAL CONSTANTS, AND SYMBOLS

A complete description of the International System of Units is given in Page, C. H. and Vigoureux, P., National Bureau of Standards, Special Publ. 330, January 1971. A good general reference for physical constants and conversion factors is Mechtly, E. A., National Aeronautics and Space Administration, SP-7012, 1964. Selected unit conversion and physical constants are given in the following table.

CONVERSION FACTORS

Density

lb/ft^3	kg/m^3	g/cm^3	mol/cm^3	Amagat
1	16.018	0.016018	7.9458×10^{-3}	178.216
0.062428	1	0.001	4.9605×10^{-4}	11.126
62.428	1,000	1	0.49605	1.1126×10^4
125.85	2,015.9	2.0159	1	2.2428×10^4
5.6111×10^{-3}	0.089881	8.9881×10^{-5}	4.4586×10^{-5}	1

Specific Volume

ft^3/lb	m^3/kg (l/g)	cm^3/g	cm^3/mol
1	0.062428	62.428	125.85
16.018	1	1,000	2,015.9
0.016018	0.001	1	2.0159
7.9458×10^{-3}	4.9605×10^{-4}	0.49605	1

Pressure

lb$_f$/in.^{2} (psi)	MPa	atm	Torr (mm Hg)	bar
1	6.8948×10^{-3}	0.068046	51.715	6.8948×10^{-2}
145.04	1	9.8692	7,500.6	10.0
14.696	0.10132	1	760.0	1.0132
0.019337	1.3332×10^{-4}	1.3158×10^{-3}	1	1.332×10^{-3}
14.504	0.1	0.98692	750.06	1

Enthalpy, Heat of Vaporization, Heat of Conversion, Specific Energies

Btu/lb	kJ/kg (J/g)	J/mol	cal/g
1	2.3244	4.6858	0.55556
0.43022	1	2.0159	0.23901
0.21341	0.49605	1	0.11856
1.8	4.1840	8.4345	1

Specific Heat, Entropy

Btu/lb-R	kJ/kg-K (J/g-K)	J/mol-K	cal/mol-K
1	4.184	8.4345	2.0159
0.23901	1	2.0159	0.48182
0.11856	0.49605	1	0.23901
0.49605	2.0755	4.184	1

Thermal Conductivity

Btu/ft-hr-R	mW/cm-K	J/s-cm-K	cal/s-cm-K
1	17.296	0.017296	0.0041338
0.057816	1	0.001	2.3901×10^{-4}
57.816	1,000	1	0.23901
241.90	4,184	4.184	1

Viscosity

lb/ft-s	kg/m-s ($N\text{-}s/m^2$)	cP (10^{-2} g/cm-s)	lb-s/ft^2 (slug/ft-s)
1	1.48816	1,488.16	0.031081
0.67197	1	1,000	0.020885
6.7197×10^{-4}	0.001	1	2.0885×10^{-5}
32.175	47.881	4.7881×10^{4}	1

Temperature

K	R	C	F
1	1.8	1	1.8

Velocity of Sound

ft/s	m/s
1	0.3048
3.2808	1

Surface Tension

lb_f/in.	N/m	dyne/cm
1	175.13	175.13×10^{3}
5.7102×10^{-3}	1	1,000
5.7102×10^{-3}	0.001	1

lb_f = pound force
lb = pound mass
g = gram
kg = kilogram mass
mol = 2.01594 gram
ft = foot
m = meter
cm = centimeter
Btu = British thermal unit
J = joules = watt second
W = watt
cal = calorie
R = absolute temperature in degrees Rankine
F = temperature in degrees Fahrenheit
K = absolute temperature in Kelvins
C = temperature in degrees Centigrade
N = Newton
MPa = megapascal (pressure)
atm = atmosphere pressure
Torr = millimeter of mercury pressure
cP = centipoise
slug = 32.174 lb
R = gas constant = 8.31434 J/mol-K, 8.31434×10^6 N-cm^3/m^2-mol-K

mol wt = 2.01594

Index

AUTHOR INDEX

A

B

C

D

E

F

G

H

I

J

K

L

M

N

O

P

R

S

T

V

W

SUBJECT INDEX

G

H

I

L

M

N

O

P

R

S

T

U

V

W

Z

CRC PUBLICATIONS OF RELATED INTEREST

*CRC HANDBOOK OF CHEMISTRY AND PHYSICS, **60th Edition***
Edited by **Robert C. Weast, Ph.D.**, Consolidated Natural Gas Co., Inc.
This Handbook is the definitive reference for chemistry and physics and maintains the tradition that has earned it the reputation as the best scientific reference in the world.

CRC HANDBOOK OF ENVIRONMENTAL CONTROL
Edited by **Richard G. Bond, M.S., M.P.H.**, and **Conrad Straub, Ph.D.**, both with the University of Minnesota.
Designed in a logical sequence to deal with the major questions on environmental control, the Handbook deals with many aspects of the environment with an emphasis on data rather than discussion.

CRC HANDBOOK SERIES IN MARINE SCIENCE
Edited by **J. Robert Moore,** Marine Science Institute, The University of Texas.
A multi-volume series that eventually will explore virtually every aspect of the marine environment, this Handbook presently deals with oceanography, marine products, mariculture, and fisheries.

CRC HANDBOOK OF MATERIALS SCIENCE
Edited by **Charles T. Lynch, Ph.D.**, Wright-Patterson AFB, Ohio.
Providing a current, readily accessible guide to the physical properties of solid state and structural materials, this Handbook is interdisciplinary in approach and content.

CRC HANDBOOK OF RADIOACTIVE NUCLIDES
Edited by **Yen Wang, M.D., D.Sc. (Med.)**, Homestead Hospital, Pittsburgh.
Information compiled from current scientific journals and other sources of scientific data were selected for this Handbook, providing a large fund of information for workers using radioactive nuclides.

CRC HANDBOOK OF TABLES FOR APPLIED ENGINEERING SCIENCE, 2nd Edition
Edited by **Ray E. Bolz, D.Eng.**, Worcester Polytechnic Institute, and **George L. Tuve, Sc.D.**, formerly with Case Institute of Technology.
Covering many fields of modern engineering, this comprehensive reference is designed to serve both the practicing engineer and the engineering student.

INDUSTRIAL CONTROL EQUIPMENT FOR GASEOUS POLLUTANTS
By **Louis Theodore, D.Eng., Sc.**, Manhattan College, and **Anthony Bounicore, M.Ch.E.**, Entoleter, Inc.
Directed toward the fundamentals and design principles of industrial control equipment for gaseous pollutants, this reference includes pertinent data with appropriate practical applications.

RECENT DEVELOPMENTS IN SEPARATION SCIENCE
Edited by **Norman N. Li, Sc.D.**, Exxon Research and Engineering Co.
Intended for both postgraduates and professionals, this multi-volume reference set discusses recent developments in the science and technology of separation and purification.

TRACE ELEMENT MEASUREMENTS AT THE COAL-FIRED STEAM PLANT
By **W. S. Lyon, Jr., B.S., M.S.**, Oak Ridge National Laboratory.
This is an examination of an exhaustive trace element balance made at the T. A. Allen Steam Plant in Memphis, Tennessee by an interdisciplinary group of scientists and technologists.

CRC CRITICAL REVIEWS IN ANALYTICAL CHEMISTRY
Edited by **Bruce H. Campbell, Ph.D.**, J. T. Baker Chemical Co.

CRC CRITICAL REVIEWS IN ENVIRONMENTAL CONTROL
Edited by **Conrad P. Straub, Ph.D.**, University of Minnesota.

CRC CRITICAL REVIEWS IN SOLID STATE SCIENCES
Edited by **Donald E. Schuele, Ph.D.**, and **Richard W. Hoffman, Ph.D.**, Case Western Reserve University.